GREEN GUIDE

Herbs

GREEN GUIDE
Herbs

BOB PRESS
ILLUSTRATED BY CHRISTINE HART-DAVIES

NEW
HOLLAND

This edition first published in the UK in 1994 by
New Holland (Publishers) Ltd
37 Connaught Street, London W2 2AZ
in association with Aura Books plc

ISBN 1 85368 230 6 (hbk)
ISBN 1 85368 287 X (pbk)

Commissioning editor: Ann Baggaley
Editor: Charlotte Fox
Designer: Paul Wood

Phototypeset by Contour Typesetters
Reproduction by Daylight Colour Ltd
Printed and bound in Singapore by
Kyodo Printing Co (Singapore) Pte Ltd

Contents

Introduction

This book is intended to introduce the reader to 150 of the most common, important and widely used herbs by describing the plants themselves, indicating which parts are used and for what purpose. It is not a herbal in the true sense, since it does not attempt to describe preparation methods, precise dosage or other information vital to the safe and successful use of herbs. Such information can be found elsewhere if required, although the reader is advised that a qualified practitioner should always be consulted before use. Despite the necessary warnings, herbs are a fascinating, useful and, above all, enjoyable area to investigate.

The relatively recent resurgence of interest in various alternatives to standard medical treatment, as well as the popularity of foreign cuisines, signals a willingness on behalf of the public to take a more active interest in herbs. Yet almost everyone already reaps the benefits of herbs and their products, whether knowingly or not. Herbs can be employed in a variety of ways, which include their use for medicinal, therapeutic and culinary purposes. In commercially produced medicines, in health and beauty products, in preserved and ready-prepared foods and in recipes prepared at home or in restaurants, herbs are an everyday part of our lives.

Medicinal herbs were originally regarded as the simple ingredients of medications, hence the old term *simples* for such preparations. Combinations of such simples formed the compound medicine which would be prescribed for sufferers. Once the only remedies available to the population at large, simples have long been a mainstay of domestic medicine. Although some have been shown to be without value or even dangerous, others have been proved effective and a surprising number have a long, unbroken history of use. As such they are as valuable today as they ever were.

Therapeutic herbs are those used in aromatherapy and other holistic preparations — they are a source of the various essential oils used in aromatherapy. Although the unique properties of essential oils were recognised some 5,000 years ago it is due to the recent interest in alternative remedies that aromatherapy is becoming an increasingly popular method of promoting good health and well-being. Essential oils are extremely concentrated and are therefore used in very small amounts in massages, aromatic baths, inhalations, compresses and various cosmetic preparations.

The use of herbs for culinary purposes began for purely practical purposes — to improve the keeping qualities of meat or, where this failed, to disguise or make more palatable the poor quality and often rotten food that was all to be had. Modern technology and health regulations such as sell-by dates have improved this state of affairs immeasurably, yet culinary herbs are still used as widely as ever. Many herbs make food more palatable by easing digestion. Indeed, they are regarded as essential ingredients in a vast array of dishes and are of great commercial importance as preservatives.

What are Herbs?

The word 'herb' can be used in a variety of ways. In a scientific sense it refers to plants which lack woody tissues but as an everyday term it also refers to plants used for culinary or medicinal purposes. This can cause confusion since many herbs in the latter sense are woody plants such as the shrubby Lavender or even trees like Willow. Even greater confusion arises over the distinction between herbs (in the everyday sense) and spices.

Herbs are plants, or parts of plants such as roots, leaves, flowers or fruits, which are used to prepare medicinal items, in therapeutic medicine, and to flavour foods. They are usually thought of as being fragrant but this is not always the case; many herbs have no particular fragrance whereas some are very strong smelling indeed. The word 'herb' also covers prepared products of herbs. Although a distinction between plant and product should strictly be made, it generally makes little difference. The dried and fragmented leaves of Oregano obviously come from the plant of the same name but in cases where the products have an appearance or a name very different from that of the plants from which they are derived, the two may not be easily linked. Mustard is derived from two different plants (*Brassica nigra* and *Sinapis alba*) while Neroli is obtained from the Seville Orange tree.

Introduction

Spices are generally regarded as the hard parts of aromatic plants, usually of tropical origin. They include roots and stems (Ginger), bark (Cinnamon), dry fruits (Cardamom) and seeds (Pepper) but some spices are actually soft parts such as flower buds (Cloves). Equally, the hard parts of many temperate plants provide herbal products and can be regarded as spices, such as Cumin seeds. Another term which can include examples of herbs and spices is 'condiment', usually defined as a seasoning which gives relish to food. This is a general term which includes inorganic substances such as salt and in the last century even vegetables such as onions and turnips were counted as condiments. In the medicinal sense even less distinction is made between herbs and spices. In addition there are a number of herbal products which fit neither category well, such as gums, resins and essential oils, and all medicinal plants tend to be referred to simply as herbs.

Perhaps the clearest difference between herbs and spices is that herbs tend to be mild-tasting and are usually more effective as flavourings when used fresh, while spices tend to be stronger and more pungent and are usually used dried. However, there is so much overlap between herbs and spices that any precise distinction is rather meaningless and plants from both categories are included in this book.

Names of Herbs

Plants, and therefore all herbs, often have two names: a popular or common name and a scientific or Latin name. Scientific names, which are the international currency of botanical nomenclature, are the more reliable and informative of the two and it is well worth becoming familiar with them and with their correct application.

A scientific name consists of two elements, showing the genus and the species to which the plant belongs. The name not only labels or identifies the plant but can provide information about the plant itself or about its relationships. Thus the Dog Rose, *Rosa canina*, and the Apothecary's Rose, *Rosa gallica* cv. *officinalis*, are separate species both belonging to the rose genus *Rosa* and therefore sharing many similar characteristics. The specific name *canina* means 'of dogs' and refers to the supposedly inferior uses of this plant compared to other roses. The name *gallica* means 'of France' and refers to the origin of that species. The cultivar name *officinalis*, which means a medicinal plant sold in shops, clearly indicates the general use to which this species was put.

Common names are more familiar and can also be informative but have various drawbacks. They vary in different languages and are not always applied consistently even within the same areas of a country. A plant may have several common names, or share the same common name with another, completely different plant. A member of the Saxifrage family, Biting Stonecrop (*Sedum acre*) is equally well known as Wall Pepper but is entirely unrelated to the Sweet Pepper which belongs to the Potato family. Cinnamon and Cassia are both members of the genus *Cinamomum* although the common names give no indication of this and the fruit of Thorn-apple (*Datura stramonium*), while spiny, is otherwise not at all apple-like as the name suggests. With herbs there is the additional complication of the products — that is the parts we think of and actually use as herbs and spices — having completely different names from the herbs from which they are derived or when a herb yields more than one product. Paprika is obtained by grinding the fruits of the Sweet Pepper and cayenne from its close relative the Chile Pepper (*Capsicum frutescens*). White pepper and black pepper are both obtained from the same plant (*Piper nigrum*). In this book, the entries are usually listed under the common name of the plant rather than the name of the product. This will enable you to more easily find the same plants in other reference works, such as guides to identification. A few entries are listed under the product name where this is better known in a herbal sense than the plant. In these cases both names are included in the text and can also be found in the index.

The Safe Use of Herbs

There is a tendency nowadays to view herbalism and especially homeopathic medicine as providing 'safe treatments', vouchsafed by hundreds, even thousands, of years of use during which time any treatments with harmful effects have been eliminated. While this view may be correct to some extent, it does not mean that herbal treatments are without risk or that they can be used with abandon, and the warnings accompanying many of the species in this book should be heeded. The situation is further complicated by the fact that some individuals will, almost inevitably, be allergic to part-icular herbs and may suffer bad reactions to treatments. Before embarking on any treatments other than simple, everyday ones such as tonics, it is essential to consult a doctor or other qualified practitioner.

Introduction

The individual entries in this book describe which parts of the herb are usually used and in what form. Dosages are not given as they can vary depending on a number of factors such as the precise condition being treated, the age and general health of the patient and even the quality and origin of the herbal material available. Again, the safest way to determine the correct dosage is to consult an expert.

Finally, a warning about identifying herbs is needed, especially when these are gathered in the wild. Obviously, it is vital to use the correct species in any herbal preparation or culinary flavouring. The required herb may have very similar relatives with which it is easily confused. In many cases these relatives are themselves innocuous, but some are harmful or even acutely poisonous. A good example is the Carrot family which contains many common herbs such as Sweet Cicely but also very poisonous plants such as Hemlock. The concise descriptions of the herbs in this book contain some, but by no means all, of the diagnostic features for each species. Unless you are totally familiar with a herb always check its identity in an authoritative botanical guide and, once again, seek expert advice.

Growing Herbs

Herbs are extremely rewarding plants to grow. Not only can they look and smell wonderful but they are also very useful. A good place for cultivating a variety of herbs is a sunlit, sheltered area with well-drained soil, although some herbs are extremely resilient and will grow in almost any conditions.

A popular style of herb garden is a formal one. This, as its name suggests, is laid out in well-defined patterns and geometric shapes, often outlined with stone or bricks. For those prepared to devote a great deal of time and attention to such a garden, the individual beds can be surrounded with dwarf hedges instead, for example of Box or Lavender.

Less formal plantings can be made by growing herbs together with flowers and vegetables in an *ad hoc* way. Many herbs are well-behaved plants which fit easily into borders and are just as attractive and useful grown in this way. It is a sensible idea to ensure that all of the herbs can be reached easily for regular harvesting.

The combinations of herbs and variations of design are endless and it is simply a matter of personal choice and, of course, the space available. Careful planning will produce a satisfying balance of colour, texture and aroma. Even in a tiny plot a small bed can be used to grow a limited but varied selection of herbs by grouping small plants together. It is not necessary, however, to have a garden at all in which to grow herbs. They can be grown almost anywhere, from hanging baskets to kitchen windowsills. Even a large pot outside the door is sufficient for a few culinary herbs to supply the kitchen.

However and wherever they are grown, remember to balance the numbers of any herbs against the extent to which they will be used. A whole bed of Thyme or Chives may be needed whereas one or two plants of Parsley will suffice. Some species, such as Ground-elder and Mint (which is said to flourish only in a house with a nagging wife) can be invasive and are best grown in a separate bed, or within a large flower pot or bucket which has had the bottom knocked out before being sunk in the soil.

Artificial chemicals should never be sprayed or applied to herbs. Few suffer from any serious pests and they can be further protected by 'companion planting' — a system in which one plant benefits from being grown in proximity to another. Aromatic herbs are known to be especially effective in this role. Common pairings include Rosemary with Sage and Coriander with Chervil.

Introduction

The precise selection of herbs grown in a garden will obviously depend on a variety of conditions such as the type and acidity of the soil but, whatever conditions prevail, not all herbs can be grown together. It is often stated that herbs in general need to be grown in full sun but obviously woodland species do best in shade and, in fact, many other herbs such as Chervil, Sweet Cicely and Valerian grow happily in partial or dappled shade. Southern herbs, especially those native to the Mediterranean region, do need plenty of sunshine, not only to grow well but to fully develop the aromatic oils which give their foliage its characteristic scent. They also prefer light, well-drained soils similar to those of their home region. They can be susceptible to frost in northern climes but with a little protection will grow successfully in many areas. Conversely, some northern species are unhappy in regions with long, dry summers and warm winters and may fail to establish. Spearmint, for example, often grows poorly in Mediterranean countries. In cooler, northern areas, frost-resistant or hardy perennial species are obviously most suitable but tender herbs can still be grown, either in cold frames or by being moved indoors during the winter.

Collecting, Harvesting and Storing Herbs

The garden is the best place to collect herbs as the plants are conveniently to hand but many herbs can also be collected in the wild. Searching for herbs in the fields and hedgerows can be fun but if you do undertake such an expedition there are some additional rules to follow. Do not collect near main roads or in areas where pesticides or other chemicals may have been used. Damage the plants as little as possible by taking only the parts you require, leaves or flowers for example, and leave the plant to continue growing. Remember that in Britain the law forbids the uprooting of any plant without the express permission of the landowner and also forbids the collecting of any part (including seeds) of some rare species. Until you are sufficiently confident in your own ability to accurately identify the plants, do collect in the company of someone who can. If there is any doubt whatsoever about the identity of a plant, leave it alone.

Other than observing these precautions, the rules for gathering herbs are the same for both wild and cultivated plants. Choose strong, healthy-looking specimens. Always handle herbs carefully and avoid bruising them; they are best carried in an open container such as a basket, never in plastic bags. The ideal season depends to some extent on the species and the parts to be gathered and some herbs, especially medicinal ones, need special treatment to preserve and store them. However, the following general guidelines apply to most herbs, including the more commonly used culinary ones.

Harvesting for Immediate Use

When using fresh herbs pick only the quantity needed and use immediately. Choose young, tender growth — a good method for perennial species is to pinch off the tips of the main shoots to encourage bushy growth from side-shoots for later use. Otherwise pick off lower leaves before they become too old to use.

Harvesting for Later Use

The best time to harvest for later use is on a sunny day after all the dew has dried out and before the sun is too strong to reduce any oil content in the plants. Again, only gather as much as can be processed quickly or the quality of the final products will be reduced. As a rule, aromatic plants are best harvested just before the flowers emerge, while herbs in which the leafy, flowering tops are used are best just after the flowers have opened. When just the flowerheads or individual flowers are being collected, pick when the flowers are well open but before they begin to fade. Seeds are collected as they ripen and it is often easiest to simply cut the whole head and handle it by the stalk. Roots are usually dug up in the autumn, after they have had time to build up reserves and are at their plumpest.

Introduction

Drying Herbs

The best place to dry herbs is in a dark, warm and airy place such as an attic or warm out-house. Good air circulation is essential to prevent them from rotting while sunlight bleaches or discolours leaves and flowers as well as reducing any oil content. The fresh herbs should be stored and any less than perfect parts discarded. Wash them only if necessary, and then briefly, and shake dry before spreading thinly on open racks or freely suspending them in bunches. Quick drying is vital to prevent loss of flavour and colouring. Artificial heat may be needed for this, especially in periods of wet weather. A number of herbs can be successfully dried in a microwave oven — experiment with the settings to obtain the best results.

Dried herbs are ready to store when the parts are brittle and break with a snap, but when drying whole plants remember that all parts do not dry at the same rate; petals dry more quickly than leaves, stems take even longer, and roots longest of all. Once completely dry, any remaining foreign bodies or unwanted parts should be picked out. Leaves can be stripped from the stems if required but should not be broken or crumbled as this causes them to lose flavour more quickly. Seeds are best shaken out of the dry heads.

Storing Herbs

Dried herbs can be stored in any convenient container as long as it is airtight. Glass jars are ideal but must be kept in a dark place as sunlight quickly removes both colour and flavour. Opaque containers are a good alternative and even tightly sealed plastic bags can be used. Herbs stored in this way will last easily until the next season's crop is ready, at which time it is best to replace them.

A suitably modern method for storing herbs is to freeze them. In this case the herbs should be fresh, having been carefully washed and dried if necessary, and packed into plastic bags. The full flavour is retained but do not use frozen herbs for garnishing food as they go limp when thawed.

The History of Herbs and Herbalism

Exactly when herbs were first used by man is unknown, though it was certainly in very ancient times, long before written records began. Precisely how the methods of preparing and applying herbs were discovered is equally a mystery. Trial and error alone seem insufficient and extremely risky given the poisonous nature of many plants and the elaborate procedures involved in many herbal preparations. The potential of some herbs must have been first recognised via their culinary use but history suggests their medicinal properties were recognised just as early. Certainly herbs were indispensable for making poor or nearly spoiled food palatable by masking the taste but it should also be pointed out that in early herbalism there was no strict division into medicinal and culinary herbs. Many plants can be used for both purposes and there were others besides, such as cosmetics, dye plants, colorants and strewing herbs which also, quite properly, came within the domain of the herbalist and apothecary.

In earliest times, herbal lore must have been passed down through generations as an exclusively oral tradition, and it is only from the time when written records appeared that we can begin to trace the history and uses of herbs with any certainty. The first records were manuscripts which were themselves often based on older knowledge. Few originals of these manuscripts have survived, but their existence can be inferred from references made to them in later works. The earliest are said to be Chinese and written about 5,000 years ago, though there are doubts about the date. Somewhat more recent records, beginning around 2800 BC are known from India, Egypt, and the Middle East and there are many references to herbs in the Bible. In Europe, probably the earliest and most influential writings were those of the Ancient Greeks such as Aristotle (384–322 BC) and his pupils who wrote discourses on the properties of plants and their use in medicine. Aristotle maintained a garden with more than 300 different species of herb.

One of the most famous of the ancient authorities on herbs was Dioscorides, a Greek residing in Rome during the first century AD and whose great work, *De materia medica*, listed over 500 different plants. Like other early writers, Dioscorides described the healing properties of plants and, although the descriptions of the plants themselves were often minimal, the illustrations were realistic and life-like.

The Romans spread what knowledge of herbalism they had throughout their sprawling empire. However, after its fall in *c* AD

Introduction

410 and the dawn of the Dark Ages in the fifth century, the flourishing herb and spice trade was sharply curtailed in northern Europe, although it did continue to some extent between southern Europe and the East. The dissemination and appreciation of herbal knowledge also suffered, though monasteries continued to be centres of learning right up to the fourteenth century. The monks maintained both libraries and herb gardens and made detailed records of herbs and their uses which greatly contributed to existing knowledge.

Some important advancements were made during the Dark Ages, introduced through the works of Arab and Persian scholars. A Persian physician, Avicenna (979–1037), discovered how to distil volatile oils from herbs and flowers thus paving the way for the use of herbs in therapeutic treatments. There was also, at this time, something of a culinary revolution, especially in Britain following the invasion of the Normans in 1066. The Normans brought with them their methods of cooking with herbs and spices and the cooking of the period would have most resembled that of the Middle East today.

The Renaissance in the fourteenth century brought advances in medicine and public hygiene and these were accompanied by advances in herbalism. In the fifteenth century the invention of printing using movable type had led to greater availability of herbals as those known previously only as manuscripts were produced in book form. The word 'herbal' means literally 'herb book' and some of the earliest survive as incunabula, the term used for books printed before 1501.

The sixteenth century was something of a golden age for herbalism, with many famous practitioners such as John Gerard, who was apothecary to King James I. He maintained a large physic garden in London where he grew his own supplies of herbs and is also credited with writing the famous *Herball* of 1597, though actually it is only a translation of an earlier, unpublished Belgian work.

Other changes which did nothing to advance herbalism also occurred during this century and two in particular had far-reaching effects. Herbs already had a long-established link with astrology but now an extreme version of this arose known as astrological botany. This school of thought attempted to associate the different herbs with particular planets and astrological signs. Diseases were also claimed to be caused and controlled by astrological influence. This rather absurd idea conveniently allowed an almost infinite range of herbs to be safely recommended for almost any condition. Astrological botany became particularly influential in seventeenth century England, where the well-known herbalist Nicholas Culpeper (1616–1654) was a principal exponent of it.

The second major change during the sixteenth century was the introduction of the so-called Doctrine of Signatures, an idea initiated by Theophrastus Bombast von Hohenheim (1493–1541), a doctor and professor at Basle who was better known as Paracelsus. He believed that the physical appearance of a plant indicated its medicinal properties. A good example is St John's-Wort, a herb used for treating wounds. According to the Doctrine, the translucent dots on the leaves signified the porosity of the skin and therefore the plant's ability to heal cuts on the human body. The reddish juice expressed from the flowers represented blood which served to confirm the herb's value in treating wounds.

This superstitious side to herbalism was not confined only to followers of astrological botany or the Doctrine of Signatures. The gathering and preparation of herbs had long involved considerable superstition and often the use of specific paraphernalia. Ancient Greek herb-gatherers were strongly advised to dig up Paeony root only at night lest they be seen by a Woodpecker which, enraged by the activity, would peck out their eyes. Such scaremongering may well have started as a means of safeguarding the herb-gatherers' trade by frightening off the uninitiated but similar stories became entrenched and were taken very seriously. A well-known example concerns the Mandrake, the root of which was thought to be in the form of a man or woman, complete with arms, legs and the ability to move.

Towards the end of the seventeenth century herbalism began to decline as the disciplines of medicine and botany started to separate. Herbals were superseded by Pharmacopoeias, which dealt with the medicinal aspects of plants, and Floras, which provided descriptions of them. Decreasing production of home-made remedies and their

Introduction

increasing replacement with proprietary and synthetic products bought at chemists' counters was also instrumental in the decline.

In the twentieth century medicine and botany more or less completely parted company and allopathy — what we think of as modern pharmaceutically based medicine — gained predominance over the older, traditional practices. As late as the beginning of this century, the bulk of allopathic medicines were still derived directly from plants but today almost all are pure, synthetic chemicals. There are certain advantages to this, such as consistency of quality, but there are also disadvantages such as undesirable side-effects.

It was as a reaction against the apparent shortcomings of allopathy that the opposing discipline of homeopathy came into being. Modern homeopathy was devised in the early nineteenth century by a German physician, Samuel Hahnemann. It uses only natural drugs and is essentially based on two principles, that like cures like, and (unlike allopathy) that a drug becomes more effective when diluted. So a drug which in very large doses causes the same symptoms in a healthy person as a disease does in someone who is ill can be used to treat that disease when administered in minute doses. As an example, a patient with a bad cold accompanied by red and watering eyes might be treated with minute doses of Onion since, in large doses, Onion produces the same symptoms in a healthy person. While supporters of homeopathy claim various successes, detractors counterclaim that these are due either to psychosomatic effects or to the replacement of unsuitable synthetic medicines with harmless but ineffective homeopathic ones. Whatever the truth of the matter, there is as yet no hard evidence that the homeopathic theory is correct.

Despite the rise of allopathy, herbalism has never completely died out, even in the developed countries. It survives everywhere in some form or other, and in many areas is still a major source of aid and comfort and will doubtless continue to be so. In the East, modern medicine and traditional herbalism have continued side by side as parallel and often complementary systems and in many under-developed and Third World countries herbal remedies are the only ones widely available. Even in countries such as Britain and the USA a revival of herb use is occurring on a broad front.

Practices such as aromatherapy while perhaps not in the strict tradition of herbalism, are nevertheless closely akin to it and share various of its aspects. Although essential oils were used thousands of years ago by ancient civilisations for perfume, flavouring, healing and embalming, it was a French chemist, Professor René-Maurice Gattefosse who originated the term 'aromatherapy'. He used essential oils to treat wounded soldiers in World War I and claimed that such oils, particularly Lavender, were extremely effective in detoxifying the body of harmful substances caused by infected wounds. Madame Maury, a pupil of Gattefosse, recognised the potential of using plant essences for skin care and developed the massage techniques now usually associated with aromatherapy. Aromatherapy today is still undergoing the process of development and research is continuing into the unique properties of essential oils and their potential for healing.

All of this is occurring in a climate of renewed public interest, as people seek a return to what they vaguely think of as a safer, more sympathetic and 'natural' approach to medicine and health in general. Yet, as some of the examples given here show, the mere fact that a particular herb was advocated by a herbalist some centuries in the past is no absolute guarantee of its efficacy.

From its very beginnings, herbalism has included treatments which were charms rather than remedies and which covered a whole range of eventualities. The leaves of Rosemary placed in a vessel of wine were said to bring good luck and a speedy sale of said wine, while for the strong of stomach there was a love potion involving Periwinkle and House-leek pounded together with earthworms. While many recipes may seem, at best, harmless and, at worst, likely to induce violent illness in the patient, they at least show an awareness of the value of prevention rather than cure, an ideal which we hold to strongly today.

Introduction

Modern research has shown, and continues to show, that although some of the herbs used by our ancestors are actually harmful others do, indeed, contain compounds which have the precise effects ascribed to these plants. So while we may safely disregard the wilder assertions of the early herbalists and take others with a pinch of salt, it seems there is still much to be learned about the value of herbs.

Glossary

Alkaloid Complex organic compounds containing nitrogen and obtained from plants. Their function in nature is unknown though they may provide protection against grazing animals. Many are poisonous, e.g. strychnine, but a large number are used medicinally as drugs, e.g. morphine and quinine

Analgesic A pain-relieving drug

Annual Plant which germinates, flowers, sets seed and dies in one year

Antiseptic A substance which counters infection by preventing the growth of bacteria

Axil The angle between a leaf and the stem

Biennial Plant which germinates and grows in the first year and flowers, sets seed and dies in the second year

Bulb An underground storage organ made up of scale-like leaves, fleshy and swollen with food reserves

Bulbil A very small bulb produced in the leaf axils or in place of some of the flowers and capable of forming a new plant when shed

Burr A fruit with spines or hooked, bristly hairs, dispersed by catching on the fur of passing animals

Calyx All the sepals of a flower

Capsule Dry fruit splitting when ripe to release seeds

Carcinogenic Capable of producing cancer

Catkin Slender inflorescence of small flowers, usually crowded together and wind-pollinated

Corolla All the petals of a flower

Disc floret A very small, tubular flower with equal lobes; typical of members of the Daisy family

Diuretic A drug which increases urine production

Essential oil The volatile oil produced by aromatic plants and providing their characteristic scent and taste

Inflorescence A group of flowers and their particular arrangement, e.g. a spike or an umbel

Involucral bract One of the bracts surrounding a head of small flowers or florets; typical of the Daisy family

Lanceolate Shaped like the blade of a spear, widest below the middle

Linear Very narrow, with parallel sides

Mucilage A substance which swells and becomes slimy in water

Narcotic A drug producing drowsiness leading to sleep and eventually unconsciousness

Opposite With a pair of leaves at each joint of the stem

Ovate Egg-shaped, widest below the middle

Palmate With lobes or leaflets spreading from a single point

Panicle A complex type of inflorescence, often large and always branched

Perennial A plant which flowers and sets seed each year; herbaceous perennials die down to ground level in winter but grow again in spring from underground storage organs such as bulbs

Perianth All the sepals and petals of a flower, especially when these are indistinguishable from one another

Photosensitivity A reaction caused when light-sensitive chemicals are exposed to the sun

Pinnate With two parallel rows of lobes or leaflets

Ray floret A very small tubular flower with one side of the apex extended into a long, petal-like strap; typical of the Daisy family

Rhizome A horizontal, underground stem, sometimes forming a storage organ

Semi-parasite A parasitic plant which obtains some of its food from another plant (the host); such plants have green pigment and their own root system and are often capable of living completely independently

Spur A projection formed by the sepals or petals of a flower

Stamen Male organ of a flower

Stolon A creeping stem, often rooting from the joints where they touch the ground

Style Elongated part of a flower's ovary bearing a sticky area (the stigma) which receives pollen

Tannins Water soluble chemical contained in some plants; they are bitter-tasting

Taproot A stout, main root, often acting as a storage organ

Trifoliate With three leaflets

Umbel A branched inflorescence, the branches of equal length and all radiating from the same point, like the spokes of an upturned umbrella; typical of the Carrot family

Vermifuge A substance used to drive out worms

Herbs

Juniper *Juniperus communis*
Small tree or shrub up to
600cm high. Prickly green
foliage of female trees studded
with green, berry-like cones
ripening blue-black with a dull
bloom in the second or third
year. Native throughout most
of the temperate northern
hemisphere. The berries yield
an antiseptic and strongly
diuretic oil used to treat
cystitis and, in aromatherapy,
for detoxification. Diluted, the
oil can be rubbed on to the
temples to reduce pain from
rheumatism and neuralgia but
the oil should be used
internally with great care, and
never during pregnancy. The
berries flavour gin and sauces.

White Willow *Salix alba*
Silvery-grey tree up to 25m
with upswept branches.
Leaves narrow, silvery-hairy
eventually becoming dull
green above. Catkins appear
with the leaves, males and
females on separate trees.
Grows beside rivers and
streams throughout most of
Europe, western and central
Asia. The fresh or dried bark
has long been used to treat
colds, aches and as a general
painkiller. Its effectiveness
was originally explained as
being due to the White
Willow's obvious ability to
grow unhindered in damp
places. Well-known nowadays
as containing the basis of
aspirin which is now produced
synthetically.

Hop *Humulus lupulus* Perennial
climber with twining stems up
to 600cm high. Opposite leaves
are large, usually with 3-5
lobes and bristly with stiff
hairs. Plants unisexual, males
with branched clusters of
flowers, females with papery
cones. Native to north
temperate regions and often
cultivated. The hops are used
to counter liver and digestive
disorders and to make a mild
sedative. A hop-filled pillow is
said to cure insomnia. Hops
are best known, however, for
their use in flavouring beer
and are grown on a large
commercial scale for the
brewing industry. Only the
fruiting heads are used.

Slippery Elm *Ulmus rubra*
Deciduous tree up to 30m
high. Leaves ovate up to
170mm long with an
asymmetric base, sharply and
doubly toothed margins and a
rough upper surface. Fruits
yellowish-green, papery.
Native to damp woods of
eastern North America. White
inner bark slimy, hence the
name of the tree. Powdered, it
makes a thick, mucilaginous
tea used internally to ease
sore throats, coughs and
ulcers and externally for
wounds and burns. Removal
of the inner bark damages or
even kills the tree and
increased demand has led to
the use of the much less
effective outer bark.

23

Herbs

Nettle *Urtica dioica* Coarse perennial up to 120cm high, covered with stinging hairs. Leaves ovate, pointed and toothed. Flowers greenish, small, forming axillary spikes. Plants either male or female. Found throughout the northern temperate regions. Nettles are used to treat a variety of conditions, from gout to dandruff. One painful herbal remedy is urtication which is the rubbing or flogging of the limbs with Nettles to relieve arthritis and rheumatism. Aerial parts rich in vitamins A and C, iron and other minerals and are added to soups and salads or made into nettle pudding or beer.

Sandalwood *Santalum album* Small evergreen tree up to about 10m high, with slender, drooping branches and smooth, grey-brown bark. Flowers dull yellow turning reddish-purple later, bell-shaped. The fruit is *c* 10mm across, dark red to black. A semi-parasite on other plants, possibly native to Indonesia but cultivated throughout tropical Asia. The heartwood provides medicinal extracts used to treat bronchitis and cystitis. An oil is extracted from it for perfumes and cosmetics. Supposedly an aphrodisiac, Sandalwood oil is thick, with a characteristic scent, and is used in aromatherapy to relieve tension and anxiety.

Common Sorrel *Rumex acetosa*
Perennial up to 100cm, with
acid-tasting foliage. Leaves
distinctive, arrow-shaped,
with backward-pointing basal
lobes. Flowers small with 3
broad inner perianth-
segments which become red
and papery in fruit; the sexes
are on separate plants.
Widespread throughout the
northern temperate regions
and cultivated in gardens.
Leaves high in vitamin C, and
oxalic acid giving them a
tangy, acid taste. Taken
continuously and over long
periods, it can cause the
formation of small stones of
calcium oxalate. A salad and
pot herb from ancient times, it
tastes metallic if cooked in an
iron pan.

Chickweed *Stellaria media*
Creeping annual up to 30cm
long. Stems much-branched,
up to 90cm bearing pairs of
oval, pointed leaves. Flowers
have deeply divided, white
petals slightly shorter than
the sepals. A fast-growing
weed native to Europe but
spread by man and now found
throughout most regions of
the world. Chickweed can be
made into an ointment or
poultice for inflamed skin,
ulcers and chilblains. The
leaves contain vitamin C and
their main culinary use is as
an addition to salads or even
boiled as a vegetable.

Herbs

Clove Pink *Dianthus caryophyllus* Perennial with woody base and stems 20–50cm high. Leaves bluish-green, narrow, in pairs. Flowers 35-40mm across, in clusters of 1-5. Petals rose-pink, shortly frilled at the margins. Native to southern Europe and North Africa but widely grown elsewhere and one of the earliest herbs cultivated in Britain. The fragrant flowers have a spicy scent and flavour similar to that of true Cloves. After removal of the bitter, narrow white claw at the base, the petals are used to flavour drinks, syrup, vinegar and salads, or candied to decorate cakes.

Monkshood *Aconitum napellus* Erect perennial up to 100cm, with paired, blackish tap-roots. Leaves palmately lobed, with the lobes themselves deeply cut. Flowers mauve or bluish with 5 petal-like sepals, the upper one forming a cowl-like hood. Found across much of Europe and northern Asia as far east as the Himalayas. All parts of this plant are poisonous, especially the roots, extracts of which were used to tip arrows. Despite this, it is still occasionally used to treat pain such as neuralgia, and for coughs, though only when prescribed by a doctor.

Golden Seal *Hydrastis canadensis*
Woodland perennial 15-30cm
high, growing in colonies.
Rhizome sends up erect
shoots, each with 2 lobed
leaves and a single flower.
Flower has greenish-white
stamens but lacks petals.
Native to North America
where it is becoming rare due
to over-collecting. A
traditional Indian remedy, it is
reportedly one of the most
used herbal products in the
United States. Tea made from
the bright yellow root is used
to treat inflammation of the
mouth, throat and other
mucous membranes, eye and
ear infections, jaundice and
other conditions.

Garlic Mustard *Alliaria petiolata*
Erect biennial up to 120cm
high. Leaves distinctive,
heart-shaped, toothed at the
margins and smelling of garlic
when bruised. Flowers white,
with 4 petals 4-6mm long, are
followed by slender fruits 6-
20mm long. Found throughout
Europe, North Africa and
western and central Asia. It is
little used medicinally but is
antiseptic and will soothe bites
and stings. The leaves of this
herb taste midly of Garlic and
can be used as a substitute in
salads and sauces.

Herbs

Horseradish *Armoracia rusticana*
Robust perennial up to 125cm,
with a stout tap-root. Leaves
dark green and glossy, stalked,
oblong to ovate with toothed
margins. Flowering stems
leafy, erect and branching.
Flowers white and 4–petalled.
Native to southern Europe
and western Asia but
cultivated and naturalised in
many temperate areas. The
root is very pungent and acrid
due to the presence of
mustard oil. Its stimulatory
and antibiotic properties are
useful for urinary infections,
gout, rheumatism and
circulatory problems. Grated
and mixed with cream it yields
the well-known sauce. Young
leaves can be added to salads.

Black Mustard *Brassica nigra*
Slender annual up to 100cm.
Leaves pinnately cut, bristly,
the terminal lobe much larger
than the others. Flowers
yellow with 4 petals 7–9mm
long. Fruits, slender-beaked
and containing dark brown
seeds, are pressed against the
stem. Widespread throughout
most temperate regions and
commonly cultivated. It
contains an antibiotic oil and
is used for poultices and foot
baths to stimulate the
circulation but can cause
blistering of the skin. The
hot-tasting condiment is
obtained from the ground
seeds, mustard powder being
a mixture of Black and White
Mustard with Saffron and
coloured with Turmeric.

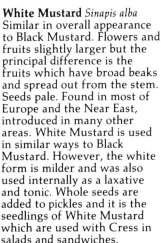

White Mustard *Sinapis alba*
Similar in overall appearance
to Black Mustard. Flowers and
fruits slightly larger but the
principal difference is the
fruits which have broad beaks
and spread out from the stem.
Seeds pale. Found in most of
Europe and the Near East,
introduced in many other
areas. White Mustard is used
in similar ways to Black
Mustard. However, the white
form is milder and was also
used internally as a laxative
and tonic. Whole seeds are
added to pickles and it is the
seedlings of White Mustard
which are used with Cress in
salads and sandwiches.

Water-cress *Rorippa nasturtium-
aquaticum* Perennial 10-60cm
high with creeping, rooting
stems growing upwards to
flower. Leaves glossy, pinnate,
with rounded leaflets. Flowers
white, small, 4-petalled. Fruits
slender with visible seeds in 2
rows on each side. Grows in
shallow, usually running
water throughout most of
Europe, North Africa and
western Asia. It can be
confused with unrelated,
poisonous species and in some
areas wild plants may harbour
a parasite – the liver fluke.
The stems and leaves have a
high vitamin and mineral
content, particularly iron, and
are used mainly in soups and
salads.

Herbs

Common Scurvy-grass
Cochlearia officinalis Biennial or
perennial 5–50cm high with
long-stalked, kidney-shaped
basal leaves in a loose rosette.
Stem leaves clasping and
fleshy. Flowers white or
occasionally lilac. Found
around the coasts of north-
west Europe and in the Alps.
The name derives from the
early use of this plant to
prevent scurvy, caused by
vitamin C deficiency, among
sailors on long voyages. The
leaves do indeed have a high
vitamin C content and were
eaten fresh. A tonic was also
made in the form of Scurvy-
grass ale; the modern version
is an infusion.

Caper *Capparis spinosa* Shrubby
perennial with straggly,
sometimes spiny branches up
to 150cm long and circular to
ovate, fleshy leaves. Flowers
white or purple-tinged, 4-
petalled, with a mass of long,
purplish stamens in the
centre. Native to parts of the
tropics and subtropics, and to
cliffs and rocky places in the
Mediterranean. Capers
contain capric acid and are
used as a condiment and in
sauces such as sauce tartare.
Only the unopened flower
buds are eaten and these must
be pickled in wine-vinegar to
bring out the characteristic
flavour.

Cinnamon *Cinnamomum zeylanicum* Small, evergreen tree up to 10m high. Leaves paired, ovate to elliptical, deeply veined and dark, shiny green. Flowers in small yellow branched clusters followed by dark purple berries. Native to Sri Lanka and the south-west coast of India, cultivated elsewhere in the east and in the West Indies. The spice is provided by the fragrant, dried inner bark of young shoots and sold either in the form of curled, papery 'quills' or ground into powder. Sometimes used to treat colds and stomach disorders. Commonly used as a sweet spice in baking and in drinks.

Cassia *Cinnamomum aromaticum* Similar to its very close relative Cinnamon though making a larger tree. Native to China and Burma, cultivated in many parts of the subtropics. Spice is obtained from 3 parts of the Cassia tree which is also known as Bastard Cinnamon. The dried inner bark is very similar to Cinnamon and is often used as a substitute although the quills are coarser both in texture and flavour. Dried leaves are used mainly in Indian cooking and the dried, unripe fruits — sometimes sold as Chinese Cassia buds — are used to flavour sweets and drinks.

Herbs

Sweet Bay *Laurus nobilis* Bushy evergreen tree reaching 20m. Leaves, wavy-edged and dotted with numerous oil glands, give off a strong spicy scent when bruised. Flowers yellowish-green, 4-petalled, with males and females on separate trees. Native to dry areas of the Mediterranean but widely grown elsewhere, both as a pot herb and as a clipped shrub. Formerly used as a strewing herb, an aid for digestion and for treating baldness. A culinary flavouring since ancient times, the leaves — dried or fresh — are an essential part of bouquet garni. Leafy branches formed the victors' wreaths of classical times.

Sassafras *Sassafras albidum* Deciduous shrub or tree 3–30m high with oval to 3-lobed, fragrant leaves developing yellow or red hues in autumn. Flowers in small, yellowish clusters appearing before the leaves. Fruits dark blue. Native to dense woodlands of eastern North America. Roots, bark, leaves and flowers were used medicinally, mainly as teas; dried leaves also formed the principal flavouring of soups and stocks. Used by indigenous peoples, and a major export during America's early colonial period, it is now banned in the United States as carcinogenic.

Star-anise *Illicium verum* Small,
white-barked evergreen shrub
or small tree 400–500cm high
with large leaves and small,
many-petalled yellow flowers.
'Fruit' consists of 8 single-
seeded pods radiating from a
central point to form a star.
Native to southern China
where it is also cultivated, and
to north-east Vietnam. Fruits
are harvested unripe and
dried. Infusions of the fruits
or seeds have been used to
treat various complaints, such
as sore throats and sickness.
The fruits are used to flavour
dishes such as beef and fish.
The oil, Oil of Anise, flavours
drinks.

Ylang-Ylang *Cananga odorata*
Evergreen tree up to 33m with
smooth, ashy bark and large,
wavy-edged leaves. Flowers
large, drooping with 6 narrow
petals *c* 75mm long. Greenish
at first, they turn yellow and
have a jasmine-like scent.
Native from tropical Asia to
Australia and cultivated
elsewhere in the Far East. The
flowers provide a heavily
scented oil with relaxing, anti-
depressant properties, which
is used to treat shock and, in
aromatherapy, for tension and
high blood pressure. Due to
the strength of the oil,
excessive or long-term use can
cause headaches and nausea.

Herbs

Opium Poppy *Papaver somniferum* Erect bluish-green annual up to 100cm. Leaves pinnately lobed. The 4 white, pink or purple petals sometimes have a dark basal patch. The large, globose capsule has holes around the rim, acting like a pepper-pot to release tiny seeds. Probably originating in the Mediterranean region, it is now widespread both as a cultivated plant and a weed. Raw opium, obtained from the milky sap of the unripe capsules, yields various medicinal drugs including morphine and codeine as well as the addictive drug heroin. The ripe seeds which contain no drugs are used in cooking.

Bloodroot *Sanguinaria canadensis* Early flowering woodland perennial 15–30cm high with blood-red roots. Flowers white, up to 50mm across, with 8–10 petals. Flowers usually appear before the leaves. Native to eastern and central North America. The fresh root was formerly used to treat coughs, asthma and lung ailments, especially by the American Indians, and it is now known to contain anti-cancer and antiseptic alkaloids. However, it is poisonous except in minute doses. Its only current commercial use is as a constituent of toothpastes and mouthwashes for combating plaque.

Biting Stonecrop *Sedum acre*
Tufted evergreen perennial 5–
12cm high. Leaves 3-6mm long,
fleshy and swollen, crowded
on short, sterile shoots,
though more widely spaced on
flowering ones. Flowers
bright yellow. Grows on dry,
alkaline soils, often on walls,
and is found throughout
Europe, northern and western
Asia, North Africa and North
America. It is slightly
poisonous and though once a
medicinal herb, taken
internally it can cause blisters
and is now used mainly as a
corn remover. However, the
dried and ground leaves of
this plant have a hot, peppery
taste and are sometimes
recommended as a seasoning.

Raspberry *Rubus idaeus*
Perennial producing erect,
woody, biennial stems up to
150cm high and armed with
weak, straight prickles. Leaves
pinnate with 5-7 leaflets
which are densely white-hairy
beneath. Flowers white,
nodding and borne in small
clusters. Native to cool
regions of Europe, northern
and central Asia but widely
cultivated. Raspberry-leaf tea
is a cure for various childhood
chills and fevers but is most
often recommended during
the later stages of pregnancy
as it tones the muscles in
preparation for childbirth.
The popular edible fruits are
used to treat kidney problems
and anaemia.

Herbs

Dog Rose *Rosa canina*
Deciduous, often scrambling
shrub up to 500cm, with stout,
hooked prickles and pink or
white flowers. Fruits 10-20mm
long, may be globose, ovoid or
elliptical. Native to Europe,
North Africa and parts of
Asia, and naturalised in North
America. Petals can be used
for perfume, the leaves to
treat wounds or as a laxative.
The fruits, called hips, contain
more vitamin C than even
citrus fruits, and those of the
Dog Rose contain most of all.
They are used to make rose-
hip syrup and the increasingly
popular rose-hip tea.

Apothecary's Rose *Rosa gallica*
Spreading, deciduous shrub
up to 100cm high with
prickles. Flowers crimson and
fragrant. Fruits bright red and
globose. Native to Europe
from Belgium southwards. It
is both a medicinal and
culinary plant and is used for
flavourings, perfumes,
powders and oils. The petals
were, and occasionally still
are, used for strewing and are
added to pot pourri. The oil
distilled from the flowers is
used in aromatherapy to
reduce tension, emotional
stress and post-natal
depression, and to cure
insomnia.

Meadowsweet *Filipendula ulmaria* Tall perennial reaching 200cm. Leaves pinnate with pairs of large, toothed leaflets interspersed with much smaller ones. Flowers creamy-white, numerous and crowded into an inflorescence up to 250mm long. They have a heavy, rather cloying scent. Found throughout most of the temperate northern hemisphere. An infusion made from the fresh flowers can be taken for all those conditions for which aspirin would normally be used. This is hardly surprising since Meadowsweet contains the chemicals which produce aspirin and, indeed, the drug is named from the old latin name for the plant, *Spiraea*.

Agrimony *Agrimonia eupatoria* Perennial with mostly basal leaves and a long flower spike up to 150cm high. Leaves pinnate with 2-3 pairs of small leaflets between each pair of large ones. Fruits crowned with hooked bristles. Native throughout Europe and extending into Asia Minor and North Africa. The green aerial parts of this plant contain a high proportion of tannins which make it valuable as a gargle and digestive tonic. Modern research also suggests that it can greatly increase blood coagulation, a property which would justify its long history as a wound treatment.

Herbs

Salad Burnet *Sanguisorba minor*
Perennial with a basal rosette
of pinnate leaves and a leafy
flowering stem 10-90cm high.
Flowers small, greenish and
tightly packed into globose or
ovoid flowerheads 10-30mm
long. They have 4 sepals but
lack petals. Usually found on
chalky grassland in most of
Europe, parts of the Middle
East and North Africa;
naturalised in North and
South America. Principally
used as an addition to salads,
the leaves have a mild
cucumber flavour. The plant
often remains green through
the winter months and was
much grown in times when
fresh salad vegetables were
scarce in winter.

Wood Avens *Geum urbanum*
Perennial 20-60cm high with
pinnate basal leaves and
deeply lobed stem leaves.
Flowers bright yellow.
Fruiting heads are burr-like
containing about 70 narrow,
hairy fruits, each tipped with
a hooked spine. Native to
most of Europe and western
Asia. The root contains the
same oil as Cloves and is
similarly antiseptic. It has also
been used as a substitute for
quinine to counter fevers and,
like Agrimony, contains
tannins which act as a
digestive tonic. On a less
practical note, it was worn by
the superstitious as a charm
against evil spirits and
venomous beasts.

Lady's-mantle *Alchemilla vulgaris* Perennial 5–45cm high with a basal rosette of leaves, their kidney-shaped to circular blades with shallow, toothed lobes. Flowers yellowish-green, tiny, and grouped into dense, branched clusters. Grows in Europe, northern Asia and eastern North America, usually on acid or neutral soils. Used to prevent internal and external bleeding, particularly menstrual disorders. The leaves are made into a tea or used as a poultice; the fresh juice is said to cure acne. Cheese made from the milk of cows grazing this herb has a distinctive flavour. The young leaves can be added to salads.

Hawthorn *Crataegus monogyna* Thorny, deciduous tree up to 18m with deeply lobed leaves 15–45mm long, masses of white flowers and dark or bright red fruits. Found throughout Europe and much of Asia. Flowers and fruits are usually used as an infusion. Hawthorn helps restore high or low blood pressure to more normal levels and improves problems brought on by ageing of the heart, arterial spasms and angina. Treatment is only effective over a period of several weeks but unlike some other drugs there is no risk of habituation. A liqueur can be made from the berries.

Herbs

Broom *Cystisus scoparius* Much-branched shrub reaching 200cm, the slender, whippy twigs green and ridged. Small, trifoliate or undivided leaves often fall very early. Flowers yellow, pea-like and numerous. Pods 25–40mm long, flattened and oblong, are hairy on the margins and black when ripe. Widespread in most of Europe. Broom contains the alkaloid sparteine which is used in cardiac treatment and obstetrics and is a strong diuretic. However, the drug's composition is variable and it should not be used except under medical supervision. The pickled flower buds were an Elizabethan culinary item.

Liquorice *Glycyrrhiza glabra* Rhizomatous perennial with erect stems up to 120cm high. Leaves pinnately divided, the leaflets sticky beneath. Flowers bluish-purple, small, pea-like, borne in axillary spikes. Native to southern Europe and western Asia, also cultivated. The roots contain glycyrrhizin, which is 50 times sweeter than sugar. When changed into an acid, glycyrrhizin is similar to human adrenal hormones and its effects are like those of cortisone. Liquorice is used to treat a wide variety of problems, including arthritis, inflammation, stomach ulcers, fevers, diphtheria and tetanus and to flavour other bitter-tasting medicines. It is also a popular sweet.

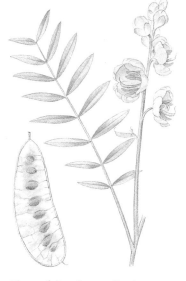

Fenugreek *Trigonella foenum-graecum* Annual 10–50cm high, the trifoliate leaves with toothed leaflets 20–50mm long. Flowers yellowish-white and pea-like, tinged violet at the base. Pods 80–140mm long, narrow, slightly curved. Probably native to south-western Asia but cultivated and widely naturalised in central and southern Europe and elsewhere. Rich in vitamins and minerals, particularly calcium. Reputedly an aphrodisiac, the seeds also contain steroid-like chemicals and in Chinese medicine are used to treat impotence and menopausal problems. The seeds are added to curries and preserves and, grown like Cress, they add a mild curry flavour to salads.

Alexandrian Senna *Cassia angustifolia* Small shrub only 50cm high. Pinnate leaves have all of the lanceolate leaflets in pairs. Flowers yellow, 5-petalled and borne in loose, erect spikes. Pods *c* 25mm wide, flattened. Native to semi-desert regions of Somalia and Yemen but cultivated in Asia. The sennas are probably the best known laxatives and Alexandrian Senna has a powerful action. The active chemicals, called anthroquinone glycosides, are found in the leaves and, particularly, in the pods. Senna is usually mixed in a syrup with other herbs and spices such as Cinnamon, Ginger or Liquorice to make it more palatable.

Herbs

Tamarind *Tamarindus indica*
Densely foliaged evergreen tree up to 30m high. Leaves 5–10cm long, pinnately divided into 10–20 pairs of closely set leaflets. Pods long and pendulous, up to 200mm, containing seeds embedded in a yellow pulp. Possibly native to tropical Africa but unknown in the wild, it is widely cultivated in India and other tropical areas. The edible pulp surrounding the seeds is used as a gentle laxative. Used locally for fevers brought on by the hot winds. It is rich in glucosides and citric, tartric and malic acids. The sharp flavour makes it a useful addition to drinks and preserves.

Nasturtium *Tropaeolum majus*
Sprawling or climbing annual with stems growing up to 200cm long. Leaves parasol-shaped. Stalk attached to the centre of the blade. Flowers orange, yellow or red, 25–60mm across with a backward-pointing spur. Fruit 3-lobed. Native to Peru but very widely cultivated, in various colour forms. All parts rich in vitamin C and antibiotic sulphur compounds which combat infection. The peppery-tasting flowers can be added to salads, the flowers and leaves used for tea, and the young pickled fruits as a substitute for Capers though they are purgative when eaten in excess.

Rose-scented Geranium
Pelargonium graveolens Perennial
up to 100cm high, with
fragrant palmately lobed
leaves. Flowers mauve or
pink, carried in loose umbels.
Native to South Africa but
cultivated in most parts of the
world. One of the ornamental
garden Geraniums, and not to
be confused with the genus
Geranium containing the
Crane's-bills. The essential oil
is used in aromatherapy
massages for relieving
premenstrual tension and
fluid retention. This species
has rose-scented foliage and
the fresh or dried leaves add
flavour to jellies, cakes and
puddings. Other species yield
a range of fragrances such as
lemon and mint.

Castor-oil-plant *Ricinus
communis* Robust annual or
spreading, shrub-like plant up
to 400cm, depending on the
climate. Leaves palmate, up to
600mm across with 5–9 lobes.
Spikes stout, containing
greenish male flowers below
prickly-looking, reddish
clusters of females. Fruits 10–
20mm, globular and spiky.
Native to the tropics but
widely grown and naturalised
in many areas. Castor oil is
obtained from the crushed
seeds and is used in engine
fuels, lubricants, paints,
varnishes and insect
repellants. It is also a mild
laxative. The seeds contain
the poison ricin and are
extremely toxic.

Herbs

Rue *Ruta graveolens* Aromatic evergreen shrub up to 45cm high, with grey-green pinnately divided leaves. Flowers 20mm across, each of the 4 yellow petals has an incurved, hooded tip. Widely grown and sometimes naturalised outside its native eastern Mediterranean region. The aerial parts yield an oil used in small doses to strengthen weak blood vessels, promote menstruation and treat colic. An ointment is used for sprains and bruises but the sap can cause a strong photosensitive reaction on contact with the skin. The plant is sometimes taken as a bitter tea.

Neroli *Citrus aurantiacum* Small evergreen tree up to 10m high, with broadly elliptical, thin but leathery leaves 7.5-10cm long. Flowers white and fragrant with 4-8 petals and numerous stamens. Fruit 75mm diameter, orange, globular, thick-skinned and bitter-tasting even when ripe. Originating in the Far East but cultivated commercially in Europe and elsewhere. Distilled from the flowers of the Seville Orange, Neroli oil is heavy and very strongly scented. It is used in aromatherapy to sooth nervous tension, to encourage sleep, and is particularly recommended for rejuvenating the skin.

Frankincense *Boswellia carterii*
Evergreen shrub or small,
papery-barked tree up to
600cm. Leaves pinnate with
spike-like clusters of small,
waxy white flowers in their
axils. Native to Somalia and
Arabia. One of the gifts of the
Magi and a major ingredient
of sacred incense,
Frankincense is extracted
from the gum-like resin of
several closely related trees of
which this species is one of
the most important. In
aromatherapy, the oil is
considered a rejuvenant,
soothing and inducing a
meditative mood to combat
anxiety, respiratory problems
and ageing skin. It is usually
blended with other oils.

Surinam Quassia-wood
Quassia amara Small tree
reaching only 600cm high.
Leaves pinnate, divided into 5
leaflets. Flowers red, tubular
and borne in clusters at the
tips of the twigs. Native to
tropical America. Both the
bark and the roots contain
bitter principles which were
used to treat dysentery, and
which are the source of the
mixer drink bitters. It should
not be confused with a green-
flowered tree from the West
Indies, also called Quassia-
wood, from which wood chips,
boiled in water, provide an
insecticide.

Herbs

Cascara Sagrada *Rhamnus purshiana* Small deciduous shrub or tree up to 12m high with pale greyish bark. Leaves 50–150mm long, prominently veined with clusters of tiny greenish flowers in the axils. Berries 8mm diameter, purplish-black when ripe. Native to western North America. The bark provides a gentle laxative which does not require repeated doses. Combined with pleasantly aromatic herbs, it is considered suitable for frail or convalescent people and is also used by vets for treating dogs. Originally used by American Indians in California, it is now exported to Europe where it has replaced treatments derived from local species.

Lime *Tilia* x *vulgaris* Tall, narrow-crowned deciduous tree up to 46m high. Leaves 60–100mm across, broad and heart-shaped, often sticky with sap. Flowers yellowish-white and fragrant hang in a cluster beneath an oblong, wing-like bract. A naturally occurring hybrid between two European species and widely planted as a street-tree. Tea made from the fragrant flowers is recommended for nervous disorders, migraines and insomnia. Lime flower tea is also used to treat colds and bronchial complaints. The inner bark is used for kidney ailments and coronary disease.

Marsh-mallow *Althaea officinalis* Densely grey-hairy perennial up to 200cm high, with large, toothed and sometimes palmately lobed leaves. Flowers, lilac-pink with shallowly notched petals 15–20mm long, form a tall spike. Native to Europe, North Africa and western Asia, introduced to North America. Marsh-mallow has a high mucilage content. The herb is used to reduce inflammation of the stomach. It also makes a gargle for throat and mouth infections. Sucking a Marsh-mallow stick is an old remedy for teething children and a sweet is made from the root.

Perforate St John's-wort *Hypericum perforatum* Rhizomatous perennial up to 100cm high, the woody-based stems with 2 raised ridges. Leaves up to 3cm, stalkless and dotted with numerous translucent glands. Flowers *c* 20mm across, yellow, with numerous stamens. Widespread in temperate regions. The foliage has an antibacterial effect and is used as a dressing for deep wounds. The red oil, extracted by steeping flowers and leaves in vegetable oil, can be used externally for neuralgia, sciatica and burns but can cause a skin reaction. A major use of the herb is in the treatment of severe depression by inducing euphoria.

Herbs

Sweet Violet *Viola odorata*
Creeping perennial up to 15cm high with a rosette of kidney-shaped leaves and long, rooting stolons. Long-stalked flowers are spurred and either dark violet or white. Native throughout much of Europe. The plant contains methyl salicylate, from which aspirin is derived and is a traditional treatment for migraines and headaches. A syrup made from the flowers, or tea from the leaves, soothes coughs and bronchial complaints. The flowers form the base for perfume and for a wine said to be good for hangovers; candied, they are used as decorative confections.

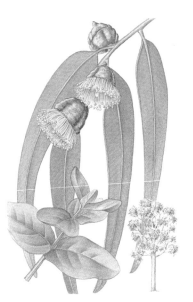

Eucalyptus *Eucalyptus globulus*
Large, fast-growing evergreen tree up to 40m high, the grey-brown bark peeling away in long strips. Juvenile leaves bluish, opposite and clasp the stem. Adult leaves dark green, alternate and drooping. Flowers are woody cups with numerous stamens but no petals or sepals. Native to Australia the Tasmanian Blue Gum is also planted in many parts of the world. Eucalyptus oil, distilled from the adult leaves, is a strong antiseptic. It is used in aromatherpy in inhalants and decongestants, as a chest rub for coughs, and as a common component of cough sweets and pastilles.

Allspice *Pimenta dioica*
Evergreen tree reaching 9m in
height, with opposite, leathery
leaves 75-150mm long. Flowers
9mm across; cream and white,
4-petalled. Males and females
are carried in clusters on
different trees. Fruits ripen
from green to dark purple.
Native to the West Indies,
Central and South America.
So named because its flavour
combines those of Cinnamon,
Clove and Nutmeg. It is used
in pickles, preserves and in
baking, and is added to many
spice mixtures. The berries
are picked unripe and dried.
The rind is the most aromatic
part and the berries are
ground immediately before
use.

Cloves *Syzigium aromaticum*
Small evergreen tree 10-15m
high with paired leaves and
clusters of bright red flowers.
These are rarely seen on
cultivated trees as it is the
unopened flower buds which,
when dried, form the cloves.
Native to the Molucca Islands
of Indonesia but also
cultivated in Zanzibar,
Madagascar and the West
Indies and always grown near
the sea. Oil of Cloves is an
analgesic used, for example, to
counter toothache. It is also
used in the perfume industry.
The flower buds are used both
whole and powdered to spice a
wide variety of culinary
dishes.

Herbs

Common Evening Primrose
Oenothera biennis Biennial with
a fleshy root and leafy stems
up to 150cm high. Flowers 45–
60mm, yellow, 4-petalled,
fragrant, and open in the
evening. Fruits long, slender
containing numerous tiny
seeds. Native to North
America but naturalised in
most of Europe. The leaves,
seeds and root are used, either
as an infusion or for
extracting an oil. The
effectiveness of this herb has
been given credence by recent
clinical testing. The oil is
successful in treating a variety
of conditions including
premenstrual syndrome and
hyperactivity, alcoholism, high
blood pressure, arthritis and
multiple sclerosis.

Witch-hazel *Hammamelis
virginiana* Shrub or small,
deciduous tree only 500cm
high, with smooth greyish
bark. Downy-hairy leaves are
widest above the middle, with
scalloped margins. Flowers
yellow, appearing in clusters
after the leaves have fallen.
Petals *c* 25mm long, strap-
shaped and very narrow.
Native to North America,
often cultivated in gardens
elsewhere. Astringent,
tannin-rich tea made from the
bark or twigs was used both
internally and externally to
maintain muscle-tone and to
treat dysentery, cholera and
other ailments. Modern,
commercially distilled extracts
and ointments are mainly
used for minor bruises and
scratches.

Nutmeg and Mace *Myristica fragrans* Evergreen tree reaching 40m high, with aromatic leaves and clusters of small, pale yellow flowers; males and females on separate trees. Fruit large and fleshy containing a kernel, the nutmeg, enclosed in a red net, the mace. Native to the Molucca Islands but cultivated in wet, seaside areas elsewhere, including the East and West Indies. Nutmeg and mace are dried and sold separately. Nutmeg is a fragrant, sweet spice used grated in cooking and as a digestive tonic, although it is toxic if taken in excess. Mace is similar but stronger and more pungent smelling.

Chinese Ginseng *Panax ginseng* Perennial up to 60cm, with a large root and a single whorl of palmate leaves at the top of the unbranched stem from which emerges a long-stalked umbel of yellowish-green flowers. A forest plant of north-east China. Most highly prized of the ginsengs, it is regarded as a unique tonic. The root contains saponins and steroids, hormone-like chemicals giving an 'adaptogenic' effect, which means returning the body to normal. Used to counter the weakening effects of age, stress or disease. Also thought to improve endurance and the ability to concentrate.

Herbs

Chervil *Anthriscus cerefolium*
Rather wiry annual up to 70cm
high, with bright green,
much-divided, pinnate leaves,
the lobes deeply cut. Flowers
small, white, in umbels. Fruits
7–10mm long including a
slender beak of *c* 4mm; narrow
and hairless in cultivated
plants. Probably native to
south-east Europe and
western Asia but widely
cultivated and naturalised
elsewhere including North
America. The leaves of this
culinary herb are used fresh.
It is a principal ingredient of
fines herbes mixtures but can be
used alone for its delicate,
slightly aniseed flavour.

Sweet Cicely *Myrrhis odorata*
Softly hairy perennial
reaching 200cm, with hollow
stems and foliage smelling
strongly of aniseed when
crushed. Leaves 2–3-times
pinnate with oblong-
lanceolate, toothed and white-
blotched lobes. Fruits 15–25mm
long, narrowly oblong and
sharply ridged. A mountain
plant from the Alps, Pyrenees,
Apennines and Balkan
mountains; cultivated and
widely naturalised elsewhere.
The leaves and aromatic seeds
have an aniseed flavour and
can be added to salads and
whipped cream or eaten on
their own. The chopped leaves
can be cooked with tart fruit
preserves and are a natural
sweetener suitable for
diabetics.

Coriander *Coriandrum sativum*
Annual with solid, ridged
stems 15–70cm high and 1–3-
times pinnate leaves. Fruits 2–
6mm, red brown and hard, the
two halves not separating
easily. Native to North Africa
and south-west Asia but
widespread elsewhere as both
a crop and a weed. An ancient
herb, known from at least
1500 BC. The leaves are
sometimes used as a garnish
or in curries but it is mostly
the ripe seeds that are used
for culinary purposes. They
are an ingredient of many
dishes, both sweet and
savoury, and add spice to
salads.

Anise *Pimpinella anisum*
Strongly aromatic annual 10–
50cm high. Lowest leaves
kidney-shaped, middle leaves
pinnate with broad lobes,
upper leaves 2–3 times pinnate
with narrow lobes. Flowers
white giving way to ovoid or
oblong, finely ridged fruits 3–
5mm long. Native to the
eastern Mediterranean and
western Asia but widely
cultivated. Anise yields an
essential oil containing the
compound anethole and is
used commercially in
toothpastes and cosmetics.
The distinctively flavoured
seeds are added to dishes such
as curries and make a relaxing
and very mildly narcotic tea
which relieves tight coughs.

Herbs

Ground-elder *Aegopodium podagraria* Stout and hairless perennial spreading by creeping rhizomes. Stems 40–100cm high, hollow bearing leaves divided into 3 leaflets; each leaflet may itself be divided into 3. Native to Europe and Asia, but introduced as a pot herb in some areas, including Britain. A traditional herb for treating arthritis, rheumatism and gout, to which the alternative name Goutweed alludes. An infusion or a poultice can be prepared from the leaves; an essence derived from the same parts is used in homeopathy. The young leafy shoots gathered just before flowering are edible.

Fennel *Foeniculum vulgare* Tall perennial up to 250cm, with a solid, polished stem. Leaves feathery, finely divided into numerous slender segments. Plant bluish-green except for yellow flowers. Native to the Mediterranean and southern Europe as far north as Britain but naturalised in many other countries. An ingredient of many proprietary cough medicines, an infusion of the seeds, or root, is taken for various minor ailments such as coughs, colic and lack of appetite. Diluted and unsweetened, it also makes a drink suitable for babies. The swollen bases of the leaf stalks can be served as a vegetable.

Dill *Anethum graveolens* Slightly
bluish, strong-smelling annual
20–50cm high, resembling
Fennel in its much divided,
feathery leaves and yellow
flowers. Fruits 5–6mm long,
dark brown, and bordered
with a pale wing; elliptical and
flattened. Probably native to
south-western Asia but
widely cultivated and
naturalised in many temperate
regions. Oil of Dill is a mild
sedative and dill water, made
by infusing bruised seeds. It
eases colic and is given to
babies. An ancient culinary
herb, Dill leaves are added to
fish dishes and cream sauces.
The stronger-tasting,
aromatic seeds are used to
spice vegetables, pickles and
vinegar.

Cumin *Cuminum cyminum*
Slender annual 10–50cm high,
the leaves divided into thread-
like lobes. Flowers white or
pink, 3–5 in each of the small,
simple umbels which form the
compound umbel. Fruits 4–
5mm long, finely ridged.
Native to North Africa and
south-western Asia, cultivated
elsewhere. Whole or ground
Cumin seeds are a common
ingredient of both Asian and
North African cuisines and
are frequently added to spice
mixtures. It is commonly used
in curries and is often added
to chicken, lamb and beef. A
traditional Indian drink
includes Cumin seeds and
Tamarind water.

Herbs

Celery *Apium graveolens*
Biennial up to 100cm high.
Lower leaves pinnate, the
upper divided into three
leaflets. Flowers greenish-
white, followed by ovoid
fruits *c* 1.5mm long. Found in
Europe, Asia and North
Africa. The cultivated Garden
Celery (var. *dulce*) is less
pungent-tasting than the wild
plants and is most commonly
used. The green leaves are
reputed to lower blood
pressure and can be added to
salads while the seeds bring
an aromatic flavour to stews
and are often used in the form
of celery salt. The swollen,
blanched leaf stalks of this
garden form are the well-
known vegetable.

Parsley *Petroselinum crispum*
Stout-rooted biennial 30–75cm
high with a solid stem and
sharply ascending branches.
Leaves 3-times pinnate, shiny
green. Flowers yellow.
Probably native to south-east
Europe or Western Asia but
cultivated and naturalised in
all temperate regions. Parsley
is strongly diuretic, increases
lactation and tones the uterus.
It must not be used during
pregnancy. The curly, frilled
leaves of some cultivated
forms are a popular culinary
garnish. More than a mere
decoration, they are a breath-
freshener recommended
against the lingering smell of
garlic and are an excellent
source of both vitamin C and
iron.

Caraway *Carum carvi* Much-branched biennial 25-60cm high with hollow, faintly grooved stems. Leaves 2-3-times pinnate with segments divided into narrow lobes. Flowers white or pink. Fruits 3-4mm long, ellipsoid and ridged. Found in temperate regions of the Old World, often introduced. An ancient herb, Caraway is cultivated on a large scale today. Various proprietary medicines for colic, flatulence and lack of appetite contain Caraway. As well as adding flavour, the seeds are one of the best natural stimulants for appetite, and are widely used in cooking and baking.

Angelica *Angelica archangelica* Perennial up to 200cm high, with stout, hollow, green stems. Flowers greenish. Fruits have broad, corky wings. Found from northern Europe and Greenland to Central Asia; introduced in North America. A tea made from seeds or dried root combats anaemia, bronchitis, asthma and — by inducing a strong dislike of alcohol — alcoholism. The fresh leaves can be added to soups, fish and stewed fruit, the seeds to flavour liqueurs such as Chartreuse. Best known for the candied young stems of confectionery. Angelica has a high sugar content and diabetics are recommended to avoid it.

Herbs

Lovage *Levisticum officinale*
Stout, strong-smelling
perennial 100–250cm high with
2–3-times pinnate leaves up to
700mm x 650mm, the lobes
deeply and irregularly
toothed. Flowers greenish-
yellow. Fruits 5–7mm long,
ridged and winged. Native to
Iran, cultivated and
naturalised in Europe and
elsewhere. The dried leaves
make a tea for fevers but it
should not be taken during
pregnancy or if suffering from
kidney disease. Plant has a
rather strong flavour
resembling that of Celery.
The leaves and seeds can be
added to many dishes,
especially vegetarian ones, and
the seeds are also used in
breads and other savouries.

Asafoetida *Ferula assa-foetida*
Yellowish, foul-smelling
perennial up to 200cm high
with stout, grooved stems and
thick roots. Leaves up to
350mm x 250mm are deeply 2–
4-times divided. Flowers
yellowish followed by fruits
12mm x 7mm. Native to Iran.
Asafoetida is a waxy gum-
resin derived from the root
and stem of this and two very
similar species. It is also used
in veterinary medicine. As a
condiment it is mainly used in
Persian, Afghan and Indian
cuisine, especially in
vegetarian dishes. The whole
plant smells strongly due to
the presence of sulphur
compounds but the smell
disappears after boiling.

Wintergreen *Gaultheria procumbens* Creeping, evergreen shrub only 15cm high with dark, shiny leaves which are thick and leathery. Flowers white, waxy, drooping and bell-shaped. Berries bright red, persisting throughout the winter. Native to northern and eastern North America. The fragrant leaves contain methyl salicylate, which is similar to aspirin and is used to treat rheumatism. It is extracted in the form of an oil and the oil content of frosted leaves (which turn purple) is thought to be higher than that of unfrosted leaves. Natural oil of wintergreen has been largely replaced by synthetic compounds.

Cowslip *Primula veris* Glandular-hairy perennial with a basal rosette of finely hairy leaves 5-15cm long. Flowers, deep yellow with orange spots at the base of the corolla-lobes, hang in a cluster at the tip of a stalk 100mm–300mm high. Native to Europe and temperate Asia, mostly on lime-rich soils. Cowslip has sedative and expectorant effects and the roots are used to treat whooping cough, bronchitis and arthritis. An ointment of the flowers treats spots and sunburn although some people may develop an allergic reaction known as primula dermatitis. The flowers are also used medicinally in various drinks.

Herbs

Jasmine *Jasminum officinale*
Deciduous or semi-evergreen
woody climber up to 10m.
Opposite leaves pinnate with
5-7 leaflets, each 10-60mm
long. Flowers usually white,
sometimes flushed purple on
the outside; fragrant and
tubular. Native to south-west
Asia but widely cultivated.
Jasmine oil is an expensive but
powerful fragrance obtained
from the flowers. It is used by
perfumers and is employed by
aromatherapists to treat
depression and respiratory
problems. The oil can also be
rubbed in to ease neuralgia.
Jasmine tea is a mild sedative
and a remedy for headaches.

Great Yellow Gentian
Gentiana lutea Stout erect
perennial up to 50-120cm, the
large, ribbed leaves opposite
and clasping; those towards
the base of the stem form a
rosette. Flowers yellow with
5-9 corolla-lobes. Confined to
the mountains of central and
southern Europe. All gentians
contain extremely bitter
principles; in this species they
are obtained from the dried
root. Once regarded as
something of a universal
panacea, the bitters are
effective against a wide
variety of digestive ailments
and stimulate production of
red blood cells. It should not
be taken during pregnancy or
if suffering from high blood
pressure.

Bogbean *Menyanthes trifoliata*
Aquatic plant of 12-35cm with leaves divided into groups of 3 leaflets and held above the water surface. Flowers, pink and white with fringed petals, are borne in spikes. Native to still water and bogs throughout most of the temperate northern hemisphere. A traditional tonic and purging herb, the leaves and rhizome contain bitter compounds similar to those found in gentians. Recommended for a variety of complaints, including anorexia, it stimulates the appetite but can cause vomiting in large doses. Also used for flavouring ales and other alcoholic drinks.

Woodruff *Galium oderatum*
Slender, fragrant perennial up to 45cm with creeping rhizomes and erect stems. Leaves usually 6-9 in each whorl, their margins with tiny, forward-pointing teeth. Flowers white and fragrant, each 4-lobed and 6mm across, form dense clusters. Found throughout much of Europe, North Africa and northern Asia. This traditional strewing herb contains coumarin, the compound which gives new-mown hay its distinctive scent. It is contained in some medicines to treat haemorrhoids and prevent thrombosis. The leaves and flowers make an excellent tea and are used to flavour wines for May-cups.

Herbs

Lady's-bedstraw *Galium verum*
Perennial up to 120cm high, with creeping stolons and much branched, 4-angled stems. Leaves very narrow, 8–12 in each whorl. Flowers 2–4mm across, bright yellow, 4-lobed forming a branched spike. Native to most of Europe and western Asia. Like its relatives Woodruff and Cleavers, Lady's-bedstraw was respectively a strewing herb and a wound herb but most of the old uses of this plant were connected with curdling milk. However, it is said to be less effective with modern milk, although it can still be used to impart a rich yellow colour to cheese.

Cleavers *Galium aparine* Rough and minutely prickly annual which scrambles through surrounding vegetation. Stems reach 180cm and are 4-angled with 6–9 leaves in each whorl. Flowers tiny, whitish and 4-lobed. Fruits 4–6mm across, burr-like, consisting of 2 fused globes covered with hooked bristles. Native throughout Europe, northern and western Asia. The dried aerial parts make an infusion used as a diuretic for cleansing the lymphatic system, reducing swollen glands and, traditionally, for cancer. It similarly helps against eczema and psoriasis and is a wash for sores and wounds. It is also recommended for treating dandruff.

Yellow-bark *Cinchona calisaya*
Evergreen tree up to 12m
high, with oval to oblong
leaves. Flowers small,
fragrant, pink and borne in
clusters. Native to the eastern
Andes; introduced to Asia
where it is grown in
plantations. The bark contains
a large number of alkaloids,
particularly quinine, which is
used for fevers and heart
problems and which is the
most effective treatment for
malaria. The alkaloids can be
toxic and as a drug quinine
should only be taken under
medical supervision. It is also
the flavouring used in tonic
water.

Lungwort *Pulmonaria officinalis*
Hairy perennial 20-30cm high
with clumps of spotted, long-
stalked leaves which enlarge
after flowering, the blades
reaching 160mm long. Flowers
pink and blue, funnel-shaped
and carried in terminal
clusters on leafy stems. Found
mainly in central and
southern Europe, northwards
to Britain and Sweden. Due to
their supposed resemblance to
lungs, an infusion of the
white-spotted leaves was
regarded as a remedy for
pulmonary complaints. This
traditional use has since been
confirmed, the leaves
containing soothing mucilage
and silica which restore
elasticity to the lungs.

Herbs

Comfrey *Symphytum officinale*
Erect, bristly perennial 50–120cm high. Leaves are 150–250mm long, the upper ones extending down the stem as wings. Flowers 12–18mm long, purple-violet, pinkish or white, tubular to bell-shaped, are carried in coiled sprays. Occurs in most of Europe. Comfrey has a reputation as a healing herb, a poultice of leaves or roots being effective for bruises, ulcers and burns. It contains allantoin which is absorbed through the skin and stimulates repair of tissue and bone. However, there is conflicting evidence about potential carcinogenic effects and this herb should not be taken internally.

Borage *Borago officinalis* Erect, bristly annual 15–70cm high with stalked basal leaves. Upper leaves stalkless and clasping the stem. Flowers 200–250mm across, blue, with 5 spreading, pointed corolla-lobes. The black stamens form a central cone. Native to southern Europe but widely cultivated and naturalised elsewhere. The herb has a long history of medicinal use, particularly as a tea for coughs and in treating depression. It contains the same active compound as Evening-primrose and is thought to work by stimulating the adrenal glands. The young leaves are added to salads, the leaves and flowers to summer drinks.

Vervain *Verbena officinalis* Erect
perennial 30–60cm with tough,
slender, 4-angled stems and
opposite, pinnately lobed
leaves 40–60mm long. Flowers
4mm across, pale pink, slightly
2-lipped and carried in long,
slender spikes. Grows from
Europe and North Africa to
the Himalayas, and has been
introduced to North America.
It is a herb with a long
tradition of magical and
medicinal use. A tea made
from the aerial parts is used to
treat nervous exhaustion,
headaches and migraine. It is
also said to be effective
against liver and gall-bladder
disorders. It should not be
taken during pregnancy.

Lemon Verbena *Lippia triphylla*
Deciduous shrub up to 8m
high in the tropics but much
less in cooler regions. Leaves
narrow, yellowish-green,
borne in whorls of 3. Flowers
pale lavender, 2-lipped and
grouped in slender, terminal
spikes. Native to South
America but widely cultivated
in the Mediterranean and
other parts of the Old World.
The aerial parts are strongly
lemon-scented when crushed
and are used as a tea to
improve digestion. It is also
said to counter depression,
lethargy, migraine and
vertigo. The essential oil
extracted from the plant is
used in perfumes and
liqueurs.

Herbs

Basil *Ocimum basilicum*
Perennial or, in cool regions,
an annual, reaching 50cm.
Opposite leaves are hairless
and slightly fleshy. Flowers
white or mauve-tinged, 2-
lipped and borne in whorls.
The whorls form a loose
spike. Several varieties,
including a purple-leaved
form, are cultivated. Native to
India but cultivated in many
parts of the world. The fresh
leaves are the principal
ingredient of pesto sauce. As
well as being served with
tomatoes, Basil is often grown
alongside them as companion
plants beneficial to growth. A
pot of Basil grown indoors
will also act an insect
repellent.

Virginian Skullcap *Scutellaria
lateriflora* Perennial up to 100cm
high, with opposite, ovate to
lanceolate, toothed leaves.
Flowers blue, 2-lipped and
borne in 1-sided, axillary
spikes. Each calyx has a
distinctive, shield-shaped flap
on the upper side. Native to
North America. Virginian
Skullcap is considered to be
one of the best sources of
nervines or nerve tonics. It
contains the sedative
scutellarin and is used in small
doses to treat epilepsy,
insomnia, depression, St
Vitus's Dance and other types
of nervous conditions. The
leaves can be made into a tea
but nowadays the herb is
usually offered in tablet form.

White Horehound *Marrubium vulgare* White-felted perennial 30–60cm high, with opposite, rounded, wrinkled leaves. Flowers white with a deeply bifid upper lip and a calyx-tube with 10 tiny, hooked teeth at the rim. Native from Europe and North Africa to central Asia. The aerial parts yield bitter principles and a volatile oil, and can be taken as a hot or cold infusion or as a syrup. The plant has been used for heart, liver, and digestive problems and as a quinine substitute for malaria but its main use is for respiratory conditions. Horehound candy is sold as cough sweets.

Balm *Melissa officinalis* Also called Lemon Balm, this 20–70cm high perennial has opposite leaves which smell strongly of lemon when bruised. Flowers 80–150mm long, pale yellow, white or pinkish, 2-lipped. Native to southern Europe, North Africa and western Asia, widely cultivated and introduced elsewhere. As a cordial and tea Balm has a long history, especially as a sedative and for treating nervous conditions and viral infections. The oil is used in aromatherapy and in massage where its antihistamine action is effective against eczema. The fresh leaves are used to flavour drinks, meats, jellies, jams and preserves.

Herbs

Motherwort *Leonurus cardiaca*
Strong-smelling perennial, 30–200cm, the opposite leaves cut into 3–7 toothed lobes which radiate from the leaf base. Flowers 8–12mm long, white or pale pink with densely hairy upper lips, are carried in compact whorls. Found in most of Europe. The main uses of this herb are for menstrual problems, and post-natal and menopausal anxiety. It is also used to treat rapid or irregular heart beat. The flowering herb makes a bitter tea and it is more normally taken as a syrup or in tablet form. It must not be used during pregnancy.

Betony *Stachys officinalis*
Perrennial 15–60cm high with sparsely leafy stems but a well-developed basal rosette of long-stalked, oblong leaves. Flower whorls are crowded into spikes. Corollas 15mm long, bright, reddish-purple and 2-lipped, the upper one flat. Native to Europe. An ancient herb, long held in high regard for its supposedly curative and protective powers, Betony has fallen into disuse. Among its many traditional uses is a poultice of fresh leaves to clean wounds, and the dried leaves, which provoke violent sneezing, to clear head colds.

Hyssop *Hyssopus officinalis*
Aromatic perennial or
miniature shrub 20-60cm high,
the stems woody at the base
and with narrow, opposite
leaves 10-50mm long. Loose
whorls of 2-lipped, blue or
violet flowers 7-12mm long are
carried in slender spikes at the
stem tips. Native to southern
Europe, North Africa and
western Asia, cultivated and
introduced elsewhere. The
volatile oil found in Hyssop
contains the same bitter
principle (marrubin) found in
White Horehound and
medicinally the two plants are
used for similar purposes.
Only small doses are required
and Hyssop should not be
used during pregnancy.

Summer Savory *Satureja
hortensis* Annual 10-25cm high
with narrow opposite leaves
10-30mm long. Flowers white,
pink or lilac and carried in
few-flowered whorls. Both
calyx and corolla are 2-lipped,
the calyx with the lower teeth
slightly longer than the upper.
Native to the Mediterranean
region, but also widely
cultivated. A culinary herb
with a strong, hot and
peppery taste. Commercially
the leaves provide a
flavouring for salami.
Domestically, they are used
mainly with vegetables and
rich meats.

Herbs

Oregano *Origanum vulgare*
Often purple-tinged, rather
woody and aromatic perennial
up to 90cm high with opposite,
stalked leaves. Flowers 4–7mm
long, white or purplish-pink,
2-lipped. Flowers are carried
in small spikes which are
themselves crowded into
terminal, flat-topped clusters.
Native to Europe, northern
and western Asia. Oregano is
little used medicinally, but the
essential oil is used in
perfumes, cosmetics and some
liqueurs. Used fresh or dried,
the leaves of Oregano (also
called Marjoram) have a
variety of culinary uses,
particularly in Italian and
Mediterranean cuisines and in
various meat products such as
sausages.

Sweet Marjoram *Origanum
majorana* Similar to, and often
confused with, its close
relative Oregano, Sweet
Marjoram is generally smaller
in all its parts, with most
leaves more or less stalkless.
Flowers have a 1-lipped calyx
deeply slit on one side, not
with 5 equal teeth as in
Oregano. Native to North
Africa and south-western
Asia, cultivated elsewhere and
naturalised in southern
Europe. As with Oregano, the
essential oil is used in
perfumes and cosmetics. More
delicately flavoured and
sweetly scented than its
relative, the fresh or dried
leaves are used in lighter
dishes with eggs, cream or
vegetables.

Herbs

Thyme *Thymus vulgaris*
Miniature aromatic shrub 10–
30cm high with erect or
spreading branches and
narrow, greyish-green,
opposite leaves. Flowers white
to pale purple, 2-lipped. Native
to the western Mediterranean
region but cultivated
elsewhere. Thyme oil has
rather limited medicinal uses,
mainly in mouthwashes and
cough medicines. Also used in
aromatherapy, some cosmetics
and toothpastes. It can be
toxic when used internally
and must not be taken during
pregnancy. A widespread and
popular culinary flavouring,
the fresh or dried leaves can
be added to almost any meat
or savoury dish.

Pennyroyal *Mentha pulegium*
Creeping, often mat-forming
perennial 10–40cm high.
Opposite leaves 8–30mm, smell
like peppermint. Flowers
small, lilac and carried in
dense whorls in axils of the
upper leaves. Found in much
of Europe and North Africa.
Although rather too strong-
smelling for many people,
Pennyroyal leaves can be used
in stews and stuffings. A
homeopathic essence prepared
from the fresh herb, or a hot
infusion, can be used for
coughs and asthmatic
problems. However, the
essential oil can be toxic and
must never be taken during
pregnancy.

Herbs

Peppermint *Mentha x piperita*
Perennial 30–80cm high,
varying from almost hairless
to grey-woolly. Leaves
lanceolate and stalked with a
pungent peppermint scent.
Flowers lilac-pink and carried
in an oblong spike. A hybrid
between the wild Water and
Spear Mints, it is widely
cultivated and naturalised.
The leaves are usually used in
a fresh state. They yield an
essential oil consisting
principally of menthol, which
has anaesthetic, anti-bacterial
and anti-inflammatory
properties. It is widely used in
inhalants, massage and
aromatherapy oils, and to
mask the taste of
pharmaceutical medicines.
Peppermint is also used as a
confectionery flavouring.

Red Bergamot *Monarda didyma*
Aromatic perennial up to
150cm high with square stems
and opposite leaves. Flowers
red, tubular and 2-lipped,
crowded together in dense
terminal clusters. Native to
North America, cultivated in
Europe. The volatile oil has a
scent like that of the
Bergamot Orange and the two
plants are used in similar
ways. The oil is used in
aromatherapy to treat
anxiety, depression and
infections. It may cause
uneven pigmentation if used
unadultered on the skin. The
leaves make the drink Oswego
tea, named after the Oswego
Indians of North America who
first used it.

Lavender *Lavandula angustifolia*
Much-branched, aromatic,
evergreen shrub up to 100cm
with narrow, opposite leaves
2–4cm long. Initially white-
hairy, they later turn green.
Flowers lavender-blue or
purplish, 2-lipped and carried
in dense, narrowly cylindrical
spikes. Native to the
Mediterranean region.
Lavender oil is an excellent
first-aid remedy for bites,
stings and burns. The
essential oil content is highest
in the flowers. It is used in
perfumes and cosmetics, and
in aromatherapy to treat
infection and stress. Dried
flowers placed in lavender
bags keep linen fresh. The
leaves and flowers are used in
herbal teas and tobacco.

Rosemary *Rosmarinus officinalis*
Aromatic, evergreen shrub up
to 200cm. Opposite leaves 15–
40mm long, dark green above
and white hairy beneath;
narrow and leathery. Flowers
10–12mm long pale blue, 2-
lipped, with 2 protruding
stamens. Native to the
Mediterranean region but
cultivated elsewhere. The
essential oil distilled from the
leaves and flowers is added to
pain-relieving liniments and
can be applied directly for
headaches. It has a reputation
as a hair tonic and is a
constituent of many
shampoos. Rosemary leaves
add flavour to a variety of
meats, especially lamb.

Herbs

Sage *Salvia officinalis* Somewhat aromatic, greyish shrub up to 60cm, with woolly branches. Opposite leaves wrinkled above and densely hairy beneath. Flowers up to 35mm long, violet-blue, pink or white and 2-lipped. Native to parts of the east and west Mediterranean but widely cultivated elsewhere. Though less used than formerly, Sage is still regarded as an effective treatment for colds, mouth and throat infections, and to combat the hot flushes of the menopause. It also improves the keeping quality of meat and processed foods and the leaves are added to various sausages, stuffings, pickles, cheeses and even honey.

Deadly Nightshade *Atropa belladonna* Shrubby-looking perennial 50–150cm high. Leaves 60–120mm long, ovate. Flowers violet-brown to greenish, bell-shaped and drooping. Fruits 10–20mm diameter, glossy-black when ripe. A woodland plant native to Europe, North Africa and parts of Asia. Widely used in proprietary medicines. All parts contain the narcotic alkaloid atropine which is used as a sedative and as an antispasmodic for paralysing parts of the nervous system. In eye-drops it dilates the pupils and is used in opthalmology. An extremely poisonous plant which must never be used without medical supervision.

Sweet Pepper *Capsicum annum*
Annual 30–90cm high with
bright green leaves. Flowers
white, drooping, with a loose
cone of bluish-yellow
stamens. Fruits up to 270mm
long, firm and fleshy, ripening
from green to yellow or bright
red. Native to tropical
America, cultivated in most
warm and even temperate
regions. The fruits of this
plant are well-known as a
fresh 'vegetable' but when
dried and ground the flesh
yields paprika. Available in
many grades, from mild- to
hot-tasting, paprika is used to
flavour a wide variety of
dishes.

Chili Pepper *Capsicum frutescens*
Tall, woody-stemmed
perennial up to 200cm. Leaves
and flowers similar to those of
the closely related Sweet
Pepper but fruits are generally
smaller, narrower, sometimes
twisted, and may be yellow,
orange or red. Native to
South America but widely
cultivated throughout the
tropics. Cayenne is the dried
and ground fruits of the Chili
Pepper. It is very warming
and is used externally to treat
muscle and nerve pain, and
internally to stimulate the
circulation and treat colds. It
has many culinary uses and is
a principal ingredient in Chili
powder and Tabasco sauce.

Herbs

Thornapple *Datura stramonium*
Annual 50-200cm high, with
coarse, wavy-toothed leaves.
Flowers 70-120mm, white to
pale-violet, fragrant and
trumpet-shaped. Seeds
contained in spiny, 4-
chambered capsules. Native to
North America but
naturalised in many parts of
the world. Thornapple is
related to both Deadly
Nightshade and Henbane. It
contains similar alkaloids, and
is likewise a narcotic. It is also
a painkiller and has been used
as an anaesthetic and to treat
Parkinson's disease. Despite
its medicinal uses, all parts of
the plant are highly poisonous
and must never be eaten.

Mandrake *Mandragora officinalis*
Rosette-forming perennial up
to 15cm high with a deep,
forked taproot and dark green
leaves up to 300mm long.
Flowers greenish-white and
bell-shaped. Fruits globose
resembling tomatoes, ripening
from green to yellow. Native
to central and south-east
Europe, rare in the wild but
cultivated in places. With its
unusual man-shaped root,
Mandrake was one of the
magical plants of the ancient
herbalists. It is, however, a
genuine medicinal herb, the
root yielding a strong
anaesthetic still used in
modern medicine as a pre-
operative drug. Like its
relative Deadly Nightshade, it
is poisonous.

Sesame *Sesamum orientale* Erect annual up to 60cm, with white, pink or mauve, trumpet-shaped flowers. Fruit is a 30mm-long capsule containing numerous shiny, ovoid seeds. Native to tropical Asia but widely cultivated in other hot areas. Sesame seeds, high in protein and oil, have a nutty flavour when cooked. They are used as a garnish or ground and added to various dishes, particularly in eastern Mediterranean cuisines. The seeds are often sprinkled on bread or biscuits. The poly-unsaturated oil expressed from the seeds is used for cooking and makes a margarine suitable for low-cholesterol diets.

Great Mullein *Verbascum thapsus* Tall biennial up to 200cm, the whole plant densely covered with white or greyish down. Flowers 12–35mm across, yellow, crowded in a long terminal spike. Upper stamens have white hairs on the filaments. Native to Europe and Asia, naturalised in North America. The leaves provide a tea which, when strained, soothes coughs and bronchial complaints. Fresh leaves make an effective compress for wounds, burns or chilblains while dried leaves are added to herbal tobacco and smoked to ease asthma. The flowers softened in olive oil can be used for ear-drops or as a liniment.

Herbs

Foxglove *Digitalis purpurea*
Biennial or perennial up to
180cm, with softly hairy basal
leaves. Bell-shaped flowers 40–
55m long form a long spike.
Petals purple, pink or white
and usually black-spotted on
the inside. Native to western
Europe. The leaves of
Foxglove yield the drug
digitalin which contains
several compounds which
affect the cardiac muscle to
increase the heartbeat.
Extremely toxic, it can cause
paralysis and sudden death if
misused. Foxglove was not
properly recognised as a
medicinal herb until the late
eighteenth century. The
similar Woolly Foxglove (*D.
lanata*) from southern Europe
has now largely replaced it in
commercial production.

Eyebright *Euphrasia rostkoviana*
Erect, branched semi-parasitic
annual up to 35cm, with
opposite, toothed leaves.
Flowers 8–12.5mm, upper lip
often lilac, lower lip white
with yellow markings. Found
throughout most of Europe
and a few adjacent areas.
There are many closely
related, similar species. As the
name suggests, the aerial
parts of this herb are used to
make a soothing eyewash, and
it is also effective against
hayfever, colds and catarrh.
An infusion can be used
externally or, diluted, taken
internally for the same
symptoms.

Greater Plaintain *Plantago major* Perennial up to 60cm, the large, strongly veined leaves forming a basal rosette. Flowers small, greenish-yellow, form a terminal spike on a stalk at least as long as the leaves. Native to Europe, North Africa and parts of Asia but now spread through most temperate regions of the world. The bruised or crushed leaves are styptic, helping to staunch bleeding, and will draw the pain from bites, stings or burns. They contain mucilage, tannins and silica, and an infusion is used to treat bronchitis, coughs and lung problems.

Pepper *Piper nigrum* Woody climber up to 600cm high with smooth, twining stems and large, thick and leathery leaves. Flowers small, greenish and petal-less, carried in long drooping spikes. Berries ripen from green to orange then red. Native to tropical Asia, but widely cultivated in the tropics. An essential condiment, pepper is one of the earliest and most valued of the eastern spices. Black peppers are the dried, unripe berries. Soaking and removing the outer skin of the unripe berries yields the milder white peppers.

Herbs

Elder *Sambucus nigra* Small, bushy tree or shrub with fragrant white flowers but foetid, unpleasant-smelling foliage. Opposite leaves pinnate, with 5-7 leaflets. Flower heads 10-24cm across, branched and flat-topped, nodding when the black berries are ripe. Native to most of Europe, North Africa and western Asia. Elder flowers are used as an infusion to treat catarrh and hayfever, as an eyewash and gargle, and to make a skin ointment. Both flowers and berries are traditionally used to make wines and cordials, and the berries are excellent in jams and pies. All other parts of the plant are poisonous.

Valerian *Valeriana officinalis* Downy perennial up to 200cm, usually unbranched, with pinnate or pinnately lobed leaves. Flowers pale pink and funnel-shaped, unequally 5-lobed and carried in a compound head made up of smaller, dense heads. Found in woods and grassy places from Europe to Japan. The roots are dried then macerated in cold water. Another of the sedative drugs, Valerian acts on the central nervous system and is good for anxiety, tension and nervous headaches. It also lowers blood pressure. If used over long periods the drug is reported to become addictive.

Boneset *Eupatorium perfoliatum*
Perennial 30–120cm high. The bases of each pair of lanceolate, wrinkled leaves are fused to encircle the stem. Flower heads with a few white or pale purple flowers are carried in flat clusters. Native to North America. Leaf tea was traditionally and widely used by Indians and early settlers to treat fevers and influenza. It was claimed to be particularly successful in the United States during the influenza epidemics of the 19th Century. Modern research suggests that the herb stimulates the immune system. It can be toxic in large doses.

Goldenrod *Solidago virgaurea*
Downy perennial up to 100cm with lanceolate to ovate leaves widest above the middle. Flower heads yellow, with both disc and ray florets, and are carried in branched spikes. Found in a variety of habitats throughout the temperate northern hemisphere. A mild diuretic, this herb is used in homeopathy and in numerous proprietary medicines for kidney and bladder disorders, as well as for arthritis and rheumatism. It can be taken as an infusion made from the aerial parts collected before the flowers fully develop.

Herbs

Yarrow *Achillea millefolium*
Erect perennial up to 60cm,
downy, aromatic and with
much-divided, dark green
leaves. Flower heads up to
10mm across in flat-topped
clusters. A plant of grassy
places, native to Europe and
western Asia, introduced
elsewhere. A traditional herb
used throughout the northern
hemisphere. Tea made from
the dried flowering plant is
used to treat colds and fevers.
The plant has styptic
properties, helping to control
both internal and external
bleeding and helps clear blood
clots. It contains at least one
toxic compound and in large
doses may cause
photosensitive skin reactions.

Elecampane *Inula helenium* Tall,
downy perennial up to 250cm,
the lower leaves stalked, the
upper stalkless and clasping
the stem. Flower heads up to
60–80mm across, florets
yellow. Native to south-
eastern Europe and western
Asia, cultivated and widely
naturalised in other temperate
regions. Tea made from the
roots is a traditional
treatment for asthma,
bronchitis, pneumonia and
whooping cough and as a
wash counters neuralgia,
sciatica and skin diseases. It
was formerly used to treat
tuberculosis. In Chinese
medicine the flowers are used
to treat some forms of cancer.
Elecampane is also used in
confectionery and to flavour
drinks.

Sunflower *Helianthus annuus*
Tall, stout annual reaching
300cm, sometimes more.
Flower heads may reach
300mm across with golden ray
florets surrounding the brown
disc. Seeds often striped black
and white. Native to North
America but cultivated
everywhere both on a
commercial scale and as a
garden ornamental. All parts
are used, for various
purposes. Tea made from the
flowers treats lung problems
and malaria, that made from
the leaves treats fever and
bites; both can cause allergic
reactions in some people.
Sunflowers are best known,
though, as a source of high
quality, edible oil, obtained by
crushing the seeds.

Scented Mayweed *Matricaria
recutita* Also called German
Chamomile. A strongly
aromatic annual up to 60cm,
with much-divided leaves.
Flower heads 30-50mm across,
the white ray florets turned
downwards. The yellow
central disc is high-domed and
hollow. Probably native to
southern and eastern Europe
and parts of Asia but a
widespread weed. This herb is
a constituent of skin
ointments and shampoos and
has an unsubstantiated
reputation as a treatment for
cancer. Tea made from the
dried flowers is used for
insomnia, hyperactivity and
anxiety. Individuals who are
allergic to Ragweeds may
suffer a similar reaction to
this herb.

Herbs

Chamomile *Chamaemelum nobile*
Also called Roman Chamomile.
A hairy, aromatic perennial
10–30cm. Similar in appearance
to Scented Mayweed but
flower heads 18–25cm across
have a conical, solid disc of
yellow florets and the white,
outer ray florets are
sometimes lacking. Native to
western Europe and North
Africa but often cultivated
and naturalised elsewhere. It
contains similar compounds
to, and is sometimes used in
place of, Scented Mayweed:
the blue oil extracted from
Chamomile is both more acrid
and less effective than that of
its close relative. Because it
forms mats, it is the species
planted for Chamomile lawns.

Tansy *Tanacetum vulgare*
Strongly aromatic perennial
up to 150cm high with
glandular, pinnately lobed
leaves. Flower heads yellow,
button-like with up to 100 in
each flat-topped cluster.
Found throughout most of
Europe and much of northern
Asia but often as an escape
from cultivation. The dried
aerial parts are a traditional
insecticidal and vermifuge
herb, formerly used both
internally and externally.
Except in homeopathy,
internal use of Tansy is now
discouraged as it is poisonous,
the essential oil especially
proving fatal if taken even in
small doses. In the United
States it is illegal to sell this
herb.

Feverfew *Tanacetum parthenium*
Yellowish-green, aromatic
perennial up to 60cm with
pinnately lobed leaves. Flower
heads 10–25mm across. Native
to the Balkan Peninsula and
western Asia but long
cultivated for medicine and
naturalised in many regions of
the world. A sedative tea
made from the leafy parts is a
traditional treatment for
arthritis, colds and cramp but
recently the plant has received
much attention as a cure for
migraine. However it should
be used with care as it can
cause an allergic reaction,
dermatitis and, particularly,
mouth sores.

Costmary *Balsamita major* Dull
green, densely hairy perennial
up to 120cm high with large
oblong, finely toothed leaves.
Flower heads 10–16mm across
carried in branched clusters.
The white ray florets are
sometimes absent, giving the
heads a button-like
appearance. Native to western
Asia but widely introduced in
Europe and North America.
Originally a brewing herb
used to flavour beers; the
leaves and flowers of
Costmary are still sometimes
used to flavour meat and
cakes, or to make a tonic tea.
It has a minty or balsam-like
scent and strong flavour, so
little is required.

Herbs

Wormwood *Artemisia absinthium*
Aromatic, woody perennial
30–90cm high. Leaves white,
silky-hairy, 2–3 pinnately
lobed. Flower heads 3–5mm
across containing only disc
florets. Found in most of
Europe though probably only
introduced in some areas, as it
is in North America.
Wormwood is a vermifuge
which is nowadays mainly
used as a tonic and digestive
aid. It formerly provided the
bitter flavouring in the
liqueur Absinthe, but it is now
banned in alcoholic drinks as it
contains the compound
thujone which irreparably
damages the central nervous
system. With the thujone
removed, however,
Wormwood is an approved
food flavouring.

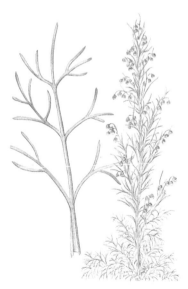

Southernwood *Artemisia
abrotanum* Strongly aromatic
shrub up to 100cm with finely
divided, gland-dotted leaves
grey-hairy beneath. Flower
heads 3–4mm across, yellowish
and button-like. Of uncertain
origin, it is widely cultivated
and naturalised, especially in
southern Europe. Closely
related to Wormwood,
Southernwood is similarly
used as a tonic, to expel
worms and, when dried, to
make a moth repellant. The
tonic aids menstruation and
should not be used during
pregnancy. The young leaves
and shoots, bitter but lemon-
tasting with a strong lemon
scent, are used for flavouring
cakes.

Tarragon *Artemisia dracunculus*
Aromatic, hairless perennial
60–120cm high, narrow leaves
mostly entire but the lower
ones 3-toothed at their tips.
Flower heads small, yellow,
globose, drooping on
downcurved stalks. Native to
southern and eastern parts of
Russia, widely cultivated and
naturalised in many other
areas. Unlike the related
Wormwood and
Southernwood, Tarragon is
used solely as a culinary herb.
Introduced into Britain in the
mid-fifteenth century.
Tarragon is used in many
sauces, *fines herbes* mixes,
marinades and preserves,
particularly in French cuisine.
It is best used fresh, or with
the flavour preserved in oil or
vinegar.

Colt's foot *Tussilago farfara*
Perennial producing flowering
stems up to 15cm high before
the large, round, shallowly
lobed and toothed leaves
appear. Leaves up to 300mm
when mature, green above,
white woolly beneath. Found
in damp waste places in most
of the temperate northern
hemisphere. A soothing tea or
syrup made from the leaves
and flowers acts on the mucus
membranes and is a
widespread cure for coughs
and bronchial problems. The
dried leaves can be smoked to
produce a similar effect and
are thought to have an
antihistamine action. The
plant is potentially toxic in
large doses.

Herbs

Marigold *Calendula officinalis*
Much-branched perennial up to 70cm with oblong to spoon-shaped, glandular downy or sparsely woolly leaves. Flower heads 40-70mm across, yellow or orange. Possibly originating in central and southern Europe but cultivated in many regions. The flowers are antiseptic and as a compress can be used for burns and ulcers. The edible petals, collected as the flowers are opening, are added to salads and stews. They can be used as a substitute for Saffron in colouring cheese, butter and cakes.

Burdock *Arctium lappa* Coarse, downy biennial reaching 150cm. Leaves broad with heart-shaped bases and solid stalks. Flower heads 30-40mm across, prickly, containing only purple disc florets. A plant of shady places in Europe and parts of western Asia. Burdock is said to purify the blood, and is also used to treat boils, measles and other skin eruptions. It is usually taken as a tea prepared from just the roots or the roots and leaves together. The seeds are used for abscesses, bites and bruises. Burdock is still used to make Dandelion and Burdock beer.

Blessed Thistle *Cnicus benedictus*
Cobweb-hairy annual 10-60cm
high with pale green,
pinnately lobed leaves up to
300mm long, the lobes
pointing backwards and
fringed with small, spiny
teeth. Stem leaves are spine-
tipped. Flower heads large
with yellow florets and
surrounded by the upper
leaves. A Mediterranean
weed, cultivated and
naturalised elsewhere.
Although slightly toxic in
large doses, causing nausea, a
weak tea made from the
flowering plant is said to be
beneficial for a wide variety of
medical problems, from colds
and migraines to jaundice,
ringworm and even deafness.
Mainly used as a digestive
tonic and appetite reviver.

Safflower *Carthamus tinctorius*
Spiny, thistle-like annual up
to 60cm high. Basal leaves
pinnately lobed but stem
leaves are undivided. Flower
heads have spiny, leaf-like
involucral bracts and
numerous yellow, orange or
red florets. Native to western
Asia but cultivated and often
naturalised in southern and
central Europe. Safflower tea
was used to treat measles but
it is principally a dye-plant.
The flowers yield two colours,
red for dyeing silk and yellow
for colouring food. It is
sometimes used as a Saffron
substitute. Nowadays the
seeds provide a source of
dietary oil which is low in
cholesterol.

Herbs

Chicory *Cichorium intybus* Blue-flowered perennial with branched stems up to 120cm high. Basal leaves pinnately lobed, stem leaves may be entire and clasp the stem. Flower-heads, containing ray florets only, are 25–40mm across and carried in small groups. Native to Europe, North Africa and western Asia, introduced into most other temperate regions. Extracts from Chicory roots have been used medicinally, mainly as diuretics and laxatives, and have a depressive effect on heart rate. The root is best known used as an additive to or even substitute for coffee. The very young, blanched leaves are a popular salad vegetable.

Dandelion *Taraxacum officinale* Variable perennial up to 50cm high, with a rosette of usually pinnately lobed leaves. Flower-heads 350–500mm, on stout, hollow stalks; only yellow ray florets are present. A widespread plant of temperate grassland and waste ground, often a weed. The leaves and fresh root make a powerfully diuretic tea while the weaker dried root can be roasted to make a coffee substitute. The whole plant is edible, the leaves and flowers rich in vitamins, especially A and C. The flowers are used for wine and the buds can be pickled like capers.

Aloe *Aloe vera* Stemless
perennial with creeping
stolons, forming clusters of
leaf-rosettes. Leaves 35–60cm,
blue-green and spiny,
sometimes tinged with red.
Flowers 25–30mm long,
drooping and cylindrical, are
carried in spikes up to 50cm. A
coastal plant native to the
Mediterranean region. Often
encountered as an ingredient
of shampoos and cosmetics,
Aloe yields a soothing
medicinal gel extracted from
the leaves. This gel stimulates
regeneration of the skin when
smeared onto cuts or burns,
and Aloe has the alternative
common name of First-aid-
plant. The juice from the cut
leaves is a very strong emetic.

Lily-of-the-Valley *Convallaria
majalis* Perennial up to 37cm
with ovate leaves arising
directly from a creeping
rhizome, their sheathing bases
forming the stem. Flowers
white or pink, fragrant and
carried in an erect, 1-sided
spike. Native to Europe and
north-east Asia but a common
garden plant. The leaves of
Lily-of-the-Valley contain
cardiac glycosides similar to
those of Foxglove, and which
have a similar effect in
strengthening and regulating
the heartbeat. It is an
alternative treatment for
cardiac patients also suffering
from high blood-pressure. All
parts of the plant are
poisonous and it must be used
only under medical
supervision.

Herbs

Chives *Allium shoenoprasum*
Tuft-forming perennial up to
50cm high with narrowly
conical bulbs less than 1cm in
diameter attached to a short
rhizome. Stem and 1–2 slender
leaves are cylindrical and
hollow. Flowers 7–15mm long,
lilac to pale purple or rarely
white, bell-shaped and
crowded in a dense umbel.
Found in most of the northern
hemisphere and widely
cultivated. Purely a culinary
herb, Chives have a milder
and much more delicate
flavour than the related
onions. Unlike onions and
Garlic, it is the finely chopped
leaves that are used, not the
small, thin bulbs.

Garlic *Allium sativum* Perennial
with stems up to 100cm high.
Bulbs 3–6cm in diameter and
composed of 5–15 small bulbs
enclosed in a papery sheath.
Leaves up to 60cm long, flat
and grass-like. The umbel
contains a few cup-shaped,
white, pink or purple flowers
but many bulbils. Native to
central Asia but widely
cultivated. It is also an ancient
medicinal herb, reducing blood
pressure, clots, sugar- and
cholesterol-levels. Its
antiseptic and antibiotic
properties are employed in
treating infected wounds and
intestinal disorders. Best
known as a culinary herb it is
used in many dishes.

Saffron *Crocus sativus* Autumn-flowering species, the grass-like leaves appearing after the flowers. Flowers lilac-purple with a yellowish throat; goblet-shaped. The prominent 3-branched style is orange. Only the large styles are used; at least 60,000 flowers are required to yield 1 pound of the spice, and it has always been very expensive. A well-known food dye, Saffron imparts a sweet scent as well as an orange-yellow colour to rice dishes, soups and cakes. Nowadays, inferior but cheaper substitutes such as Turmeric and Safflower are often used instead.

Couch Grass *Elytrigia repens* Dull- to bluish-green perennial with tough, far-creeping rhizomes. Flowering stems up to 120cm high, hairless and stiffly erect. Flowering spikes 50–200mm long, slender, unbranched and composed of paired spikelets. Found throughout much of the northern hemisphere, often as a pernicious weed. The pale rhizomes are made into a tea or decoction. A diuretic rich in minerals and vitamins A and B, it has antibiotic properties and is used to treat incontinence, kidney stones and other problems of the urinary tract. In Africa it is regarded as an antidote to arrow poisons.

Herbs

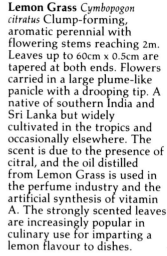

Lemon Grass *Cymbopogon citratus* Clump-forming, aromatic perennial with flowering stems reaching 2m. Leaves up to 60cm x 0.5cm are tapered at both ends. Flowers carried in a large plume-like panicle with a drooping tip. A native of southern India and Sri Lanka but widely cultivated in the tropics and occasionally elsewhere. The scent is due to the presence of citral, and the oil distilled from Lemon Grass is used in the perfume industry and the artificial synthesis of vitamin A. The strongly scented leaves are increasingly popular in culinary use for imparting a lemon flavour to dishes.

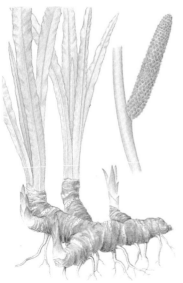

Sweet Flag *Acorus calamus* Perennial up to 125cm, the linear leaves with wavy margins and aromatic when crushed. Flowers yellowish-green, tiny and packed into a compact, up-curved spike. Native to southern and eastern Asia but naturalised in Europe and North America. The rhizome has various medicinal uses. Chewed raw or made into a tea, it reduces stomach acidity and, because it stimulates digestive secretions and therefore appetite, is recommended for anorexia nervosa. It is also used to help smokers to give up tobacco. Some strains contain a carcinogen but in others the chemical is absent.

Ginger *Zingiber officinale*
Perennial up to 100cm with
short, fleshy and knobbly,
branched rhizomes. Stems
erect with 2 ranks of leaves,
the sheathing leaf-stalks
forming the stem itself.
Flowers, yellow or white with
a purple lower lip, form dense
spikes. Native to tropical
south-east Asia but also
cultivated in Africa and the
Caribbean. It makes a
warming and mildly
stimulating treatment for the
circulation and for bronchial
complaints. The rhizome,
either fresh or dried, peeled or
not, is a culinary spice; the
crystallised stem is also eaten.

Turmeric *Curcuma domestica*
Perennial up to 100cm similar
to its close relative Ginger but
the leaves are all basal, their
sheathing stalks not, or
scarcely, forming a stem.
Rhizomes *c* 25mm diameter,
yellowish on the outside and
deep orange within. Probably
native to India, cultivated
there and in other parts of the
tropics. The boiled, dried and
powdered rhizome has a
characteristic pungent smell.
Widely used as both a
flavouring and as a colouring
agent for imparting a brilliant
yellow to a variety of dishes, it
is an essential ingredient in
curries.

Herbs

Cardamom *Elletaria cardamomum*
Perennial up to 350cm. Related to Ginger, it also has thick, fleshy rhizomes and very tall sterile stems formed by the sheathing leaf-stalks. Flowering stems leafless and spreading. Flowers white with blue and yellow markings and a single protruding stamen. Capsules 10-20mm long, ovoid, greenish-grey, are harvested unripe and dried whole. Each contains 3-4 brown seeds. Native to hills of southern India but also cultivated in Sri Lanka and parts of Central America. Dried, ground seeds are used to flavour sweet and savoury dishes and for spicing wine but are best known as an ingredient in curries.

Vanilla *Vanilla planifolia*
Evergreen, climbing orchid, up to 30m, with fleshy leaves. Flowers 40-70mm long, greenish-yellow with an inrolled, orange-striped lower lip. Pods up to 200mm long are fragrant when ripe and contain numerous tiny seeds. Native to Central America but cultivated throughout the tropics, principally in Madagascar. Outside its native range, the flowers must be hand-pollinated. First used by the Aztecs to flavour chocolate drinks. The unique flavour is due to vanillin crystals on the surface of the pods. Pods are still used for culinary flavouring, despite the availability of synthetic essence.

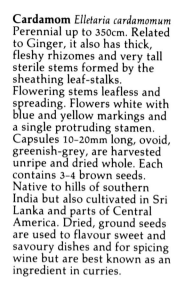

Organisations

American Herb Society 300 Massachusetts Avenue, Boston, MA 02115, USA.

The Australian Herb Society Incorporated, PO Box 110, Mapleton 4560, Australia.

Herb Federation of New Zealand, PO Box 007, Christchurch, New Zealand.

The Herb Society, PO Box 415, London SW1P 2HE, England.

The Herb Society of America, 9019 Kirtland – Chardon Road, Mentor, Ohio 44060, USA.

The National Herbalists Association of Australia, Suite 14, 249 Kingsgrove Road, Kingsgrove, New South Wales 2208, Australia.

The National Institute of Medical Herbalists, 41 Hatherly Road, Winchester, Hants, England.

Further Reading

Blackwell, W.H., *Poisonous and Medicinal Plants*. Prentice Hall, New Jersey, 1990.

British Pharmacopoeia, vol I–II. HMSO, London, 1980.

Duke, J.A. and Ayensu, E.S., *Medicinal Plants of China*, vol I–II. Reference Publications, Algonac, Michigan, 1985.

Foster, S. and Duke, A., *A Field Guide to Medicinal Plants: Eastern and Central North America*. Houghton Mifflin, Boston, 1990.

Jain, S.K. and DeFilipps, R.A., *Medicinal Plants of India*, vol I–II. Reference Publications, Agonac, Michigan, 1991.

Launert, E., *Edible and Medicinal Plants of Britain and Northern Europe*. Hamlyn, London, 1981.

Lou, Z., *et al* (eds), *Colour Atlas of Chinese Traditional Drugs*, vol I. Science Press, Beijing, 1987.

Mabey, R., *The New Age Herbalist*. Collier Books, New York, 1988.

Phillips, R. and Foy, N., *Herbs*. Pan, London, 1990.

Stace, C.A., *New Flora of the British Isles*. Cambridge University Press, Cambridge, 1991.

Stuart, M., (ed.) *The Encyclopedia of Herbs and Herbalism*, Orbis, London, 1985.

Tutin, T.G. *et al* (eds), *Flora Europea*, vols I–V. Cambridge

Herb Gardens and Collections to Visit

The Butser Ancient Farm Research Project, Petersfield, Hampshire.

Cambridge Botanic Garden, University Botanic Garden, Cambridge.

Chelsea Physic Garden, London SW1.

Gerard House, 736 Christchurch Road, Boscombe, Bournemouth, Hampshire.

Hatfield House, Hatfield, Hertfordshire.

Ledsham Herb Garden, University of Liverpool Botanic Gardens, Ness, Cheshire.

Michelham Priory Physic Garden, Upper Docker, Nr Hailsham, East Sussex.

The Museum of Garden History, St Mary-at-Lambeth, London SE1.

The Royal Botanic Gardens, The Queen's Garden at Kew Palace, Kew, Richmond, Surrey.

Royal Botanic Garden, Inverleith Row, Edinburgh.

The Royal Horticultural Society Gardens, Wisley, Guildford, Surrey.

The Tudor Garden Museum, St Michael's Square, Southampton.

Index

Index

Index

Index